PHYSICS

OF

SAILING

PHYSICS

OF

SAILING

JOHN KIMBALL

UNIVERSITY OF ALBANY
NEW YORK, U.S.A.

CRC Press
Taylor & Francis Group
Boca Raton   London   New York

CRC Press is an imprint of the
Taylor & Francis Group, an **informa** business

CRC Press
Taylor & Francis Group
6000 Broken Sound Parkway NW, Suite 300
Boca Raton, FL 33487-2742

© 2010 by Taylor and Francis Group, LLC
CRC Press is an imprint of Taylor & Francis Group, an Informa business

No claim to original U.S. Government works

Printed in the United States of America on acid-free paper
10 9 8 7 6 5 4 3 2 1

International Standard Book Number: 978-1-4200-7376-8 (Paperback)

Visit the Taylor & Francis Web site at
http://www.taylorandfrancis.com

and the CRC Press Web site at
http://www.crcpress.com

Dedication

To Marlene

Contents

Preface

One can only describe what one knows, so most of the sailing examples presented here are based on small sailboats. This prejudice is reflected in topics ignored. For example, there is surely engaging physics in offshore navigation and global positioning systems. My lack of personal experience with these mean I would have no fresh ideas to offer. One can also notice a bias toward smaller boats by the relative lack of attention given to keels and an interest in capsizing, which is one of my specialties.

Bias is also evident in the choice of physics topics. Much of fluid mechanics requires numerical work and the acceptance of complex ideas as matters of faith. Advanced applications like "vortex sheets" cannot be found here, but the foundations of fluid mechanics, which give the overall picture, are included. Anyone planning to perform accurate calculations of lift and drag must look elsewhere.

Some of the simplifications presented here are a bit extreme. A graph of wind's force on a sail and a sketch of a high-pressure weather system were both stylized with perfect circle constructions. The ice-boat speed diagram was approximated by a double circle. More realistic approaches would give better accuracy, but circles have a special appeal for me.

Mathematics is the language of physics, and some physics related to sailing is unavoidable, complicated, and mathematical. Surprisingly, many people who love sailing do not share my affection for equations.

Although most of the equations presented here can be easily skipped, the occasional boxed formula probably deserves a cursory glace.

Some derivations should be ignored by people who are in a hurry to just get the result. These sections are shaded.

I find some physical ideas associated with sailing to be particularly appealing and elegant. This is the reason that the dimensional analysis of turbulence and the structure of wakes get extra attention. Most material presented here can be found elsewhere. An exception is the scaling model for wake drag, which has not withstood the test of time.

Physics Facts

If one does not worry about circular definitions, all of classical physics follows from

$$\vec{F} = m\vec{a}$$

The total force on an object is $\vec{F}$. The object's mass is m, and the acceleration $\vec{a}$ is the rate of change in velocity.

An important application to sailing occurs when the speed is constant. Then, the acceleration a vanishes, and the total force must be zero. Often the total force is the sum of forces from the wind and water, so these forces must be equal and opposite when there is no acceleration.

For rotation, the analogous formula is

$$\tau = I\alpha$$

The total torque is τ, I is the moment of inertia, and α is the angular acceleration. If a boat is not tipping over, the angular acceleration is zero. This means the total torque must vanish. This occurs when the buoyancy cancels the torque of the wind and water.

The kinetic energy of an object is

$$KE = \frac{1}{2}mV^2$$

Here, V is the speed of the object, and m is its mass. The wind, the sailboat, and waves all have kinetic energy.

The power delivered to an object changes in its kinetic energy.

$$Power = Force \times Speed$$

A sailboat can move quickly over the water because of the power supplied by the wind.

Newton's laws applied to fluids yields the Navier–Stokes equation, which is simplified to the Euler equation when viscosity is ignored. Viscosity is the fluid analogue to friction.

Acknowledgments

Many generous people made this book possible. Vicky Woods, Debbie Kennedy, Hunter Currin, and Stéphane Caron donated pictures. Sally Snowden was especially helpful. She supplied a large number of photographs for me to look through. David Liguori provided technical and photographic knowledge and advice. Stewart Swift, Steven Olson, Phil Erner, and T. S. Kuan suggested improvements. Asim Mubeen helped with data analysis and the construction of figures, and Hasan Mahmood also helped with figures. CheHwi Chong, Steven Olson, and Robert Geer did surface roughness measurements. Scott Miller and David Fitzgarrald were the source of wind velocity and wave height measurements. Louisa Watrous negotiated permission to use the Annie photograph. Special thanks are due to the University at Albany for its support.

1

DEPART, DEPART FROM SOLID EARTH

1.1 Why Sailing, Why Physics, Why Both?

Sailing is not a good career choice. As W. S. Gilbert said,

> Stick close to your desks and never go to sea,
> And you all may be rulers of the Queen's Navee!

Prince Henry the Navigator knew this long before there was an H.M.S. *Pinafore*. This "explorer" who died in 1460, is often credited with extending Portugal's domain along the west African coast and developing a better sailing ship, the caravel. But Henry never went to sea.

Coleridge's "The Rime of the Ancient Mariner" reminds us that sailing can be uncomfortable,

> The ice was here, the ice was there,
> The ice was all around:
> It cracked and growled, and roared and howled,
> Like noises in a swound!

and boring,

> Down dropt the breeze, the sails dropt down,
> 'Twas sad as sad could be;
> And we did speak only to break
> The silence of the sea!

Despite its unprofitable, uncomfortable, and boring aspects, sailing still offers the sailor a mini-adventure that is rare in this age of sloth. An outstanding narrative of the rich rewards of sailing is Joshua Slocum's *Sailing Alone Around the World* (1899). Writer Arthur Ransome's critique of this extraordinary work, "Boys who do not like this book ought to be drowned at once," is inappropriate only because girls are not treated equally.

Every sailor can choose his own level of adventure, be it daysailing or a single-handed circumnavigation. Not every sailor would answer Sir Ernest Shackleton's famous newspaper ad for his 1914 Antarctic expedition: "Men Wanted for Hazardous Journey. Small wages, bitter cold, long months of darkness, constant danger, safe return doubtful. Honour and recognition in case of success." Sir Ernest can be forgiven for not inverting women too.

Faced with a sailing challenge, even villains become heroes. Odysseus and other legendary sailors would be thrown in jail today. Although Benedict Arnold is not a hero to Americans, he held off the British at Valcour Island in a naval "strife of pigmies for the prize of a continent." The unfairly notorious Captain Bligh exhibited remarkable skill in his 1789 forced sailing (and rowing) trip across the Pacific. Bligh and 18 loyal crew members were stuffed onto an open boat only 7 m long. With insufficient food and water, a sextant and a pocket watch but no charts or compass, he navigated 6700 km in 47 d to safety in Timor.

Sailing must have magical powers. If it can elevate Odysseus, Benedict Arnold, and Caption Bligh to positions of honor, think what it can do for you.

The appeal of physics is equally hard to explain. For some, finding the correct explanation of familiar or exotic phenomena offers greater exhilaration than a successful day of sailing. Although the core of physics has a special elegance, much of day-to-day science lacks the seductive atmosphere of profundity. This is certainly the case for the physics of sailing, which is dominated by numerical calculations and enmeshed in the cumbersome apparatus and intimidating mathematics of fluid mechanics.

Sailing appears to have a special appeal for those interested in science. Nobel Prize winners Albert Einstein and William Lawrence Bragg are high-profile examples. Scientific interests make sailing attractive, but scientific skills do not always translate into superior sailing ability. Albert Einstein enjoyed sailing, but he could not be called a skilled or careful sailor. He refused to wear a life jacket even though he never learned to swim.

You don't have to master the Navier–Stokes equation to sail fast, and time on the water does not improve your math skills. Nonetheless, many sailors are curious about how a sailboat works. In many ways, sailing is much more complicated than one would expect. Believe it or not, the messy diagrams and abundant formulas that follow are an attempt to make the physics of sailing comprehensible.

Only one bit of advice is offered. Scientists have a secret. They usually skip the math.

1.2 Origins

1.2.1 Egypt

Records of sailing are nearly as old as civilization. The earliest known depiction of a boat under sail appears on an Egyptian clay pot from around 3,100 BC. Being more than 5000 years old, the image is not very clear, but it roughly resembles the sketch in Figure 1.1.

The central role of sailing in ancient Egypt is seen in the two hieroglyphs of Figure 1.2. They mean "travel south" and "travel north." Not surprisingly, they can also be interpreted as "fare upstream" and "fare downstream."

The Nile was the highway of ancient Egypt. Sailing south in the prevailing north wind and drifting north with the Nile's current were the preferred means of travel (with the help of rowing and towing). The hieroglyphs eloquently describe both the direction and preferred mode of travel.

Communication is a key marker of civilization, and sailing was essential for this communication in Egypt. One can speculate that sailing was one reason Egyptian dynastic rule lasted for millennia.

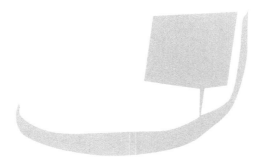

Figure 1.1 A sketch of the earliest known depiction of a sailing craft.

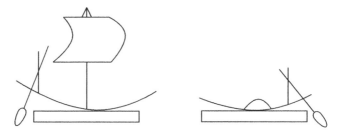

Figure 1.2   Egyptian hieroglyphs meaning "travel south" and "travel north."

Thanks to ancient religious burial customs and an extraordinarily dry climate, a ritual boat from ancient Egypt has been remarkably well preserved. More than 4,000 years ago, King Cheops (Khufu), second pharaoh of the Fourth Dynasty of the Old Kingdom, had a ritual vessel constructed. This Khufu ship was buried at the foot of the Great Pyramid of Giza where it lay, dry, disassembled and undisturbed, until 1954. The reconstructed Khufu ship shown in Figure 1.3 is now in a museum near the Giza pyramid. This boat shows no signs of sails. It may never have been in water, but it does show remarkable boat construction skills from ancient times.

Figure 1.3   The 4,000-year-old reconstructed Khufu ship in Egypt.

1.2.2 The First Sailors

Egypt's apparent historic primacy may only be the result of a climate that so effectively preserved records. There are other ancient traces of sailing. Fragments interpreted as boat parts have been found in Kuwait that date from earlier than 5000 BC. Fragments from a Turkish site in the Euphrates Valley are from about 3800 BC. A ship with mast, forestay, and backstay is depicted on a Syrian seal from around 1800 BC. Sailing ships are on Minoan seals dating from around 2000 BC. Seals from Bahrain from around 2000 BC also may show sails. A drawing from around 2000 BC in India may indicate sails, but it is not clear.

Remarkably, Australia provides some of the earliest evidence of humans living outside of Africa. How did the ancient Africans of roughly 50,000 years ago make the trip of more than 10,000 km to Australia? They could have walked most of the distance, but even with the much lower sea level of an ice-capped world, Asia and Australia were separated by water. The route is unknown, but island hopping would still require significant ocean travel. The longest step was around 200 km, which is a long way to paddle or drift. One can imagine that some rudimentary form of sailing played a role in this most ancient of all known explorations, but no one knows. Of course, rudimentary sailing can be nothing more than common sense. Gilgamesh, in what has been called the "oldest story in the world," used his shirt as a sail in his quest to find the secret of immortality.

1.2.3 Polynesia

Another extraordinary discovery era that surely did rely on sailing was the settling of the Pacific by the Polynesian people, starting around 1500 BC. Doubled sailing canoes depicted in Figure 1.4 are the likely candidates for the craft that took these people to New Zealand and the islands that sparsely populate the Pacific Ocean. Sailing these flimsy-looking sailboats, the Polynesians managed journeys that few of us would attempt today. Most impressive is the discovery around 440 AD of Easter Island, also called Rapa Nui, which means "navel of the world." Easter Island is 3000 km from anything significant (the Marquesas). Even tiny Pitcairn Island (*Mutiny on the Bounty*) is 1800 km to the west.

Figure 1.4   Petroglyph of unknown age from Easter Island and a drawing of what the actual sailing vessel may have looked like. (Drawing © by Herb Kane, with permission.)

How was the discovery accomplished? Perhaps fishermen wandered off course or chased fish long distances. Or they may have been blown to distant seas by storms. The exploration could have been planned. Prevailing winds are from east to west, so one could sail east to explore when winds were reversed and still be assured of easy return in the prevailing breeze.

Still, it is hard to understand the success. The likelihood of the occasional west wind decreases as one sails east toward Easter Island. The Polynesian craft can (and could?) sail to windward, but the sailing angles are not good, requiring a 4 to 1 ratio for distance traveled compared to distance to windward. Sailing into waves of the open Pacific Ocean would have been very hard on the canoes. Perhaps there was an alternative path. The Polynesians might have found the west winds from 35° to 50° south latitude. But weather this far south is notoriously dangerous. Another unlikely alternative is that climate could have been different 1, 600 years ago, or an unusually strong El Nino–Southern Oscillation or some other weather anomaly could have temporarily changed wind patterns so the voyages of discovery could be accomplished.

Granted that long-distance travel was somehow possible, how could people without compasses or maps travel thousands of kilometers to find an island barely 30 km across? Clouds can form above isolated islands and sometimes the cloud formations extend downwind. Terns and other sea birds sometimes cluster near islands. Even so, the discovery of Easter Island seems miraculous.

The Polynesian boats and their construction are as impressive as the journeys they made. By examining present-day Polynesian sailing canoes and scanty historical records, one can speculate on the structure and construction of these early sailboats. The two hulls were attached with crossbeams and a deck could be added. Then the double canoes could travel greater distances and carry more cargo. There were paddles, but large distances meant sailing was the essential means of transportation.

Like the Egyptians, the Polynesians had no nails, so boats were essentially tied together. The canoes were built using tools of stone, bone, and coral. The canoe hulls were gouged from tree trunks with adzes or made from planks sewn together with twisted and braided coconut fibers. Caulk was made from tree sap. The sails were woven from coconut leaves.

The result was impressive. Polynesian canoes, at least those of more than 1,000 years after the discovery of Easter Island, were fast: Around 1773, One of Captain Cook's crew on the H.M.S *Endeavour* estimated that a Tongan canoe could sail "Three miles to our two."

1.2.4 China

The western world has traditionally been unaware of Chinese contributions to science and technology in general, and to sailing in particular. For example, the Egyptian dynasties and the early Polynesians did not have rudders. The first depiction of a rudder in Europe is on a church carving of 1180 AD. But the Chinese invented the rudder 1,000 years earlier, probably in the first century AD.

It is generally accepted that the Chinese invented the first compass, dating before 80 AD. Curiously, there is also a much older Central American artifact from the Early Formative Olmec period of 1400–1000 BC. It, too, may have been a compass. The Chinese probably sailed to the coast of India prior to 1000 AD with the help of a navigational compass.

By 200 BC, Chinese were building ships the size of those used by Columbus. In the period 200–300 AD, the Chinese developed boats with multiple masts rigged fore and aft. They introduced full-length battens into their sails, resulting in a more efficient sail shape, easier handling, and greater resistance to tearing.

The first reference to centerboards and leeboards was Chinese. It dates from 759 AD.

By the 1300s, a variety of Chinese boats were constructed with watertight compartments to minimize the possibility of sinking and leeboards that reduce sideslip and make progress to windward more efficient. All of these innovations took place long before they appeared in Europe.

In the early 1400s, the Chinese imperial fleet comprised hundreds of ships much larger than those of the West. Voyages of exploration led by Zheng He, the "Three-Jewel Eunuch," were extensive, but a speculation that China discovered the New World in 1421 is almost certainly wrong. A change in China's politics in 1424 ended the most active aspects of this era of exploration and trade.

1.2.5 Speculations

Chicken bones dating from 1400 AD were found in Chile. These bones had a genetic mutation also seen in chickens native to Samoa and Tonga. Since chickens can't fly very far, one can speculate that the Polynesians made it all the way to South America.

It is commonly believed that North America was settled by migration through Siberia. However, very old Clovis arrowheads found on the east coast of the United States appear similar to stone tools found in France. A speculation that Europeans sailed along the southern edge ice age ice sheets around 13,000 years ago to North America is feasible. A lack of confirming genetic evidence means any early European sailors to North America were minority immigrants.

The northern passage through the Bering Strait was blocked by glaciers until about 13,500 years ago. However, there is (unreliable?) evidence of humans in the Americas long before that time. If early migrants did make it to the Americas, they would have come a different way. Perhaps using sailing craft, they skirted the shores of

Japan-Siberia-Alaska-Canada-California, and so on. Kelp beds may have helped make seas safer and provided anchors.

1.3 There's Much More

There are more than 1,000 reasons to love sailing. Find a sailor and you will get plenty of stories. You will hear how sailors enjoy the most beautiful sunsets, survive blistering heat, and witness waves taller than basketball players. But it is far better to take up sailing yourself so you can tell your own stories.

There are 10,000 important facts in the history of sailing. The few speculations from antiquity I chose to describe are only a preface to an increasingly complex story. An expert can tell you the other 9,990 intriguing details. A European viewpoint tells you that "yacht" has a Dutch origin, and sailing characteristics of the Spanish Armada changed the history of the world. But sailing from the earliest times has developed worldwide. Sailboats in the Middle East, the Orient, and many other areas have separate histories. The character of different people can be seen in their boats. It is hard to confuse a Chinese junk, a Mideast dhow, and a European clipper.

There are 100,000 connections between science and sailing, but it is an exaggeration to say that physics can really explain how sailboats work. Much of sailing is related to the complex motion of air and water. The mysteries of fluid mechanics blended with the mystery of sailing can present daunting questions. Despite this, basic physics can take one a long way in the description of sailboat motion, which is the goal of the following chapters.

2
DOWNWIND—THE EASY DIRECTION

Sailing with the wind is surely the oldest and simplest type of sailing. In its primitive form, it is hardly sailing at all. When the wind is behind, standing up in a canoe or mounting a small tree on a raft could be considered downwind sailing.

It is logical to wonder how fast a boat can go when sailing downwind. Physics gives clues about how to make faster sailing craft for both upwind and downwind sailing. The starting point is a general discussion of speed. Sailboat speeds are determined by the wind's force and water's opposing force. Newton's simple characterization of the forces provides reasonable estimates of sailboat speeds. A more complicated description of the forces is needed for upwind sailing.

2.1 Speed

The appeal and challenge of sailing arises, in part, from the enormous variation of conditions that occur as the wind speed varies from calm to a gale. A typical sailboat's speed is comparable to the wind speed, so calms produce bored and frustrated sailors, while a really strong wind results in panic.

A *knot* is the historical nautical speed unit. In the "old days" (perhaps all the way back to the Netherlands in the 1500s), a series of knots were tied on a rope, separated by about 15 m. (A meter is a little more than a yard.) The number of knots paid out to a fixed point in the water in about 30 s gave the boat's speed, U, in knots. Wind speeds, W, are also often quoted in knots.

One can use knots, kilometers per hour, miles per hour, furlongs per fortnight, or any other unit to characterize speed. The quaint units of the American system and other historical oddities can be avoided by always specifying speed in meters per second. It is easy to estimate boat speed in meters/second using an analogy of the old knots measurement. If a man overboard at the bow of a boat can be hauled up at the stern after 10 s, and if the boat is 10 m long, then you know the boat speed is 1 m/s. It must be a light wind. On the other hand, if this unfortunate crew zips past the stern after one second, you know the boat speed is 10 m/s. It is very windy.

A qualitative characterization of wind speed is the Beaufort scale, developed around 1800. This scale places wind speeds into about a dozen categories. A modern refinement of the Beaufort scale relates the Beaufort numbers, B, to the wind speed, W.

$$W = 0.836 \cdot B \cdot \sqrt{B} \, \frac{\text{meters}}{\text{second}} \tag{2.1}$$

The Beaufort number $B = 5$ is called Fresh Breeze. Many small boat sailors find this wind, which can produce white-capped waves, to be more than sufficient. They become nervous at much greater wind speeds. Others, particularly sailboarders, happily deal with more wind. Table 2.1 compares Fresh Breeze wind speeds in different units. It can be used as a conversion table for those more at ease with alternate speed units. For example, doubling the speed in meters/second gives the approximate speed in knots.

Desirable wind speeds (for sailors) are similar to the speeds of human locomotion. Normal walking at a little more than 1 m/s corresponds to a minimum usable wind for sailing. The world's fastest sprinters struggle mightily to obtain the Fresh Breeze speed of 10 m/s. Ice boats are a special case. Ice boat speeds of 30 m/s or more, corresponding to automobile highway speeds, can be obtained.

For simple calculations, Fresh Breeze will be taken as 10 m/s and Gentle Breeze (3 on the Beaufort scale) to be half that speed, or 5 m/s.

Table 2.1 A comparison of speed measures

BEAUFORT NUMBER	METERS/SECOND	KILOMETERS/HOUR	MILES/HOUR	KNOTS
5	9.35	33.6	20.9	18.2

2.2 Forces

Fluid mechanics explains why even a modest increase in wind speed makes sailing much more adventurous. The "quadratic approximation" is the key.

2.2.1 Quadratic Approximation

The force a moving fluid (air or water) exerts on an object (the sailboat) is proportional to the square of the fluid velocity (relative to the sailboat).

As Figure 2.1 shows, increasing the speed by a mere factor of 5 multiplies forces by 25. This relationship explains why anyone can trim a sail in "light air," but it takes a surprising amount of strength to manage the same sail in a Fresh Breeze.

It may appear that someone forgot to label the units in Figure 2.1. This is not an oversight. The parabolic shape of the quadratic approximation has a universal quality, which means it doesn't matter if the wind speed extends from zero to 10 m/s or to 50 m/s, and it doesn't matter if the force is applied to a giant sail or a tiny sailboat hull. It doesn't matter if it is the force of the wind or the force of the water. The shape of the curve is always the same. Only the scales on the axes would differ.

Isaac Newton was the first to obtain a plausible justification of the quadratic approximation by inventing the "impact theory" of fluid resistance. Volume 2 of Newton's *Principia* derives concepts of fluid mechanics from basic principles of mechanics, which he was the first to formulate.

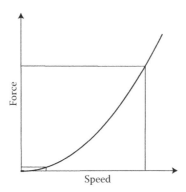

Figure 2.1   The quadratic relation between speed (of the wind or water) and force (on the sail or the boat). Inserted lines show that a 25-fold increase in the force is produce by a 5-fold increase in speed.

To test his theory, Newton dropped inflated hog's bladders and glass balls filled with air and mercury from the top of St. Paul's Cathedral in London. The quadratic approximation was confirmed. If Professor Newton plotted his data, the results would resemble Figure 2.1

Even in the "old days," military concerns have intersected science. The relation between force and fluid velocity continues to attract considerable attention because the air resistance of projectile motion has obvious military applications.

2.2.2 Newton's Impact Theory

Today, Newton's impact theory is regarded as naïve. However, in Newton's time, much was unknown about the nature of matter. So when Newton postulated "corpuscles" of fluid hitting a surface, he used this term because atoms and molecules were unknown. With his knowledge of mechanics, he could reason that each corpuscle exerts a force proportional to its velocity. The flux, or rate at which corpuscles hit a surface, is also proportional to the velocity. The quadratic approximation is a consequence of multiplying the two velocity terms. The following fills in some details of the quadratic approximation for wind on a sail. It is in the spirit of Newton's impact theory.

A boat is sailing downwind with its sails extended to capture the wind, as shown in Figure 2.2. Visualize each air molecule striking a sail to be a tiny bullet with mass, m, and an average speed, V (with respect to the boat).

A sail stops the molecular bullets (Newton's corpuscles). To do so, the sail must exert a force on each molecule for a short impact time, τ. The stopping molecules exert an equal and opposite force on the sail, pushing it forward.

For each molecule, one can multiply Newton's law, $f = ma$, by the impact time τ to get

$$\tau \cdot f = \tau \cdot ma \qquad (2.2)$$

The acceleration a multiplied by time is the wind speed ($V = a\tau$). Thus,

$$f = \frac{mV}{\tau} \qquad (2.3)$$

Figure 2.2 A Flying Scot sailboat sailing downwind. (Photograph by Sally Snowden. With permission.)

Equation 2.3 shows, as expected, that the single-molecule force f is proportional to the wind speed. However, during the impact time, N molecules hit the sail and all the forces add to give the total force $F_D = N \cdot f$. To find N, multiply the number of molecules per unit volume by the volume of air that hits the sail in time τ. As is shown in Figure 2.3, this volume is the sail area A multiplied by the distance the air moves during the impact time, which is $x = V \cdot \tau$.

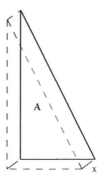

Figure 2.3   The sail area, A, multiplied by the distance x the wind travels in the time τ gives the volume of air that hits the sail in this time.

Combining Equations 2.2 and 2.3 gives

$$N = A(sail) \cdot [V\tau] \cdot (Number / Volume) \qquad (2.4)$$

Multiplying the single molecule force f of Equation 2.4 by N gives the total force.

$$F_D(wind) = m \cdot (Number / Volume) \cdot A(sail) \cdot V^2 \qquad (2.5)$$

The number density (*Number/Volume*) of air is not an easy quantity to measure. Surely Newton had no idea how small a molecule was and how many molecules were in a cubic meter of air. However, multiplying the number density by the mass of each molecule gives the mass density of the air, denoted $\rho(air)$. Today, the mass density of air is easily measured, and $\rho(air) \cong 1.25 \ \text{kg} / \text{m}^3$.

Thus, the force of the wind is expressed in measurable quantities.

$$F_D(wind) = \rho(air) \cdot A(sail) \cdot V^2 \qquad (2.6)$$

It is certainly no surprise that the force is proportional to the sail area $A(sail)$ and the density of the air $\rho(air)$. The subscript D, appended to the force, F, stands for "drag," which is a force in the direction of the fluid motion. Later "lift" will be encountered with a subscript L.

The impact theory is not restricted to air and sails. Any moving fluid (such as water) exerts a force on an object (such as a boat). The formula is essentially the same, except $\rho(air) \rightarrow \rho(water)$, $A(sail) \rightarrow A(hull)$, and the wind speed is replaced by the boat speed $V \rightarrow U$.

2.2.3 *Refinements*

Newton was well aware of the limitations of his impact theory. His model postulated a fluid that was "thin" so interactions between the corpuscles could be ignored. Newton never claimed any fluid to be thin and explicitly stated that water was not thin. Nonetheless, the impact theory was applied without careful scrutiny, due in part to Newton's fame. Later, Leonard Euler, competing father and son Johann and Daniel Bernoulli, and others rejected the impact theory. They recognized the complexity of fluid motion. One obvious modification is illustrated in Figure 2.4.

Surprisingly, even though Newton's impact theory is flawed, it remains a useful qualitative guide. To move from the impact theory

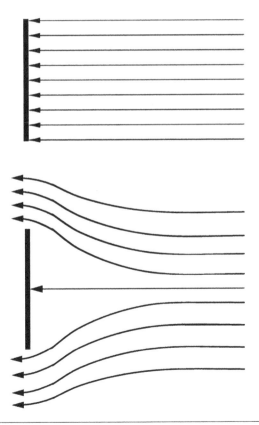

Figure 2.4   The upper diagram assumes fluid flow is stopped by an object, as in Newton's impact theory. The more realistic picture below shows fluid deflected and slowed by the object.

to the physical result, the force of Equation 2.6 is written in a modified form that has become standard notation.

$$F_D = \frac{C_D}{2}\rho A V^2 \qquad (2.7)$$

The additional "drag coefficient" C_D summarizes a multitude of corrections to the impact theory. Sailors are interested in even small changes in the drag coefficient because increased drag from the sail produces faster downwind sailing.

The "reasonable" approximation $C_D(sail) \cong 4/3$ taken from experiments allows us to obtain estimates of downwind sail forces. A measured drag coefficient, C_D, which is close to the impact theory value is partly a coincidence. Streamlined objects like centerboards have a much smaller C_D. Cup-shaped objects like spinnakers produce larger drag coefficients. An anemometer made from cone-shaped objects attached to an axis works because the drag coefficient of the pointed side is considerably smaller than the drag coefficient of the cupped side. For flat objects like a sail perpendicular to the wind, C_D does not change much with fluid speed. The limited variation of C_D is evidence that the quadratic approximation is a reasonable starting point.

A comment on the mathematics and the equal sign

Sailing physics involves a lot of formulas. Equations of special significance are placed in a box. For example, $F_D = C_D \rho A V^2/2$ (Equation 2.7) gets a box because it determines sailboat speed. Various forms of this equation appear again and again, for upwind and downwind sailboat characterizations. Even though Equation 2.7 is important, one can skip its justification in Equations 2.2–2.5. This is indicated by the shading of steps leading to the final result.

Some formulas describing the physics of sailing are accurate, but many are only reasonable guesses. Newton's formula $F = ma$ is written with an (=) sign because it is an exact (or essentially exact) statement. $F_D = C_D \rho A V^2/2$ has an equal sign because it is the definition of the drag coefficient C_D. An accurate approximation, such as the acceleration of gravity $g = 9.8$ m/(s)^2, is given honorary status with an equal sign. On the other hand, $C_D(sail) \cong 4/3$ is written with the "approximately equal" ($\cong$) sign because it is not very accurate. A sail's drag coefficient could be more or less than 4/3, but the estimate is not wrong by more than a factor of 2. If an estimate is so questionable that it could easily be wrong by a factor of 2 or more, the "roughly equal" ($\approx$) is used. Proportional quantities are indicated by ($\propto$). For example, the drag force is proportional to the drag coefficient ($F_D \propto C_D$) even when other quantities like the sail area are not considered. Of course, this is illegal math. How many ($\equiv$)'s must be multiplied together before one gets a ($\approx$)?

2.3 Boatspeed

The quadratic approximation yields uncharacteristically simple results for downwind sailing. Actually, the results are not that simple, but sailing in other directions is much more complicated. Although sailing downwind is relatively easy to describe, it is not necessarily a fast or exciting direction to sail.

Before proceeding to an answer, be warned that there are many reasons for caution and skepticism. The sailboat hull lives in the complicated interface between air and water. Light sailboats can rise up and plane over the water. Heavier boats have their speed limited by the generation of a wake, which is described in Chapter 8. Since these modifications are ignored for now, the results that follow come closer to reality for heavier boats sailing in light winds where skipping over the water's surface and wake generation are both relatively unimportant.

2.3.1 Apparent Wind Speed, V

Anyone who has taken a sailboat ride quickly notices that the wind appears to nearly vanish as a sailboat changes from sailing upwind to downwind. Three different speeds make this observation precise. They are (1) the "true wind speed," W, which is the wind speed relative to the water, (2) the "apparent wind speed," V, which is the wind speed as observed by someone on the moving sailboat, (3) the "boat speed," U, which is also relative to the water. (W for *wind*, V for *viewed* wind, and U for the speed *you* are going.) The apparent wind speed, V (not the true wind speed W), determines the wind force $F_D(wind)$. When sailing downwind, the boat's speed subtracts from the wind speed, so

$$V = W - U \qquad (2.8)$$

When the water is moving, one must make sure to measure all the speeds in the above formula with respect to the water. Someone sitting on the shore next to a river could feel no wind at all even though sailboats in the river could be moving along smartly, enjoying the relative motion of the air with respect to the water. In many situations, water speed and direction of motion vary with location. Clever planning is required to take full advantage of the complicated patterns of water currents.

2.3.2 Downwind Speed Ratio, S_0

The downwind speed ratio, S_0, compares the boat speed to the *apparent* wind speed.

$$S_0 = \frac{U}{V} \qquad (2.9)$$

Slower cruising boats are characterized by $S_0 < 1$. Light sailboats with big sails can have a downwind speed ratio greater than unity. Under ideal conditions, iceboats can have downwind speed ratios that are much greater than unity.

A sailor is normally interested in the boat speed as compared to the true wind speed, W, not the apparent wind speed, V. The ratio U/W can be expressed in terms of the downwind speed ratio using the Equations 2.8 and 2.9 ($V = W - U$ and $S_0 = U/V$). The result is

$$U = \frac{S_0}{1 + S_0} W \qquad (2.10)$$

Equation 2.10 makes good sense. The fraction preceding W is always less than unity because you can't sail faster than the wind when the wind is from behind. A graph of U/W as a function of S_0 is shown in Figure 2.5. For a typical sailboat, the downwind speed ratio is $S_0 \cong 1$, meaning the downwind sailing speed is about half the true wind speed, or about 5 m/s in a Fresh Breeze.

2.3.3 Calculating the Downwind Speed Ratio

The downwind speed ratio, S_0, is the key to sailboat speed. So how big is S_0 and how can it be increased? Common sense tells us that a light boat with a big sail will be faster than a tubby boat with a stubby sail. The following calculation of S_0 validates common sense and makes it quantitative.

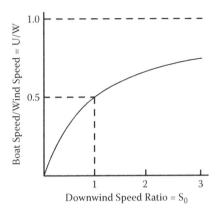

Figure 2.5   The dependence of the ratio $U/W = $ (*boat speed*)/(*true wind speed*) on the downwind speed ratio S_0.

Call the wind force on the sail $F_D(wind)$. The opposing water force on the hull, including a rudder and keel or centerboard, is $F_D(water)$. No acceleration means the opposing wind and water forces are equal.

$$F_D(wind) = F_D(water) \tag{2.11}$$

The quadratic approximation as expressed by Equation 2.7 provides expression for the opposing forces $F_D(wind)$ and $F_D(water)$

$$F_D(wind) \cong \frac{1}{2} C_D(sail) \cdot A(sail) \cdot \rho(air) \cdot V^2$$
$$\tag{2.12}$$
$$F_D(water) \cong \frac{1}{2} C_D(hull) \cdot A(hull) \cdot \rho(water) \cdot U^2$$

Equating these forces and remembering that $S_0 = U/V$ from Equation 2.9 gives an expression for the downwind speed ratio

$$S_0^2 \cong \frac{C_D(sail) A(sail) \rho(air)}{C_D(hull) A(hull) \rho(water)} \tag{2.13}$$

Equation 2.13 is sufficiently complicated to be a bit discouraging, especially because the underwater cross section $A(hull)$ is not simple to estimate. Because the goal of this calculation is insight rather than precision, $A(hull)$ is estimated using an ancient result.

2.3.3.1 Archimedes Principle "The mass of the displaced water is equal to the mass of the boat."

The boat mass $m(boat)$, which includes the mass of the crew and all the extra equipment brought on board, is the product of the water density $\rho(water)$ and the volume of the displaced water. For a streamlined hull, the volume is roughly half the boat length L multiplied by the cross-sectional area $A(hull)$. Thus, Archimedes gives

$$A(hull) \cdot \frac{L}{2} \cdot \rho(water) \cong m(boat) \tag{2.14}$$

The approximation of Equation 2.14 for the hull cross section gives a more practical expression for S_0.

The downwind speed ratio becomes

$$S_0^2 \cong \frac{C_D(sail)}{C_D(hull)} \frac{\rho(air) \cdot A(sail) \cdot L}{2m(boat)} \tag{2.15}$$

Equation 2.15 is a basic result that summarizes the common sense of sailboat speed. It is no surprise that the downwind speed ratio is larger for longer boats with bigger sails. It is no surprise that S_0 is smaller for a heavier boat. However, because these boat properties determine the *square* of S_0, their influence on boat speed is not as large as one might imagine. For example, doubling the sail area only increases S_0 by about 40%. Quantitative estimates of S_0 and the resulting sailboat speeds are given in Section 2.6.

In principle, Equation 2.15 tells one how to engineer a faster sailboat. To achieve maximum speed, sailboat designers attempt to make $C_D(sail)$ as large as possible and the counterbalancing hull drag coefficient $C_D(hull)$ should be as small as possible. Practical considerations, such as stability, limit the ratio $C_D(sail)/C_D(hull)$.

It is curious that S_0 of Equation 2.15 depends on the density of air but not on the density of water. Although it's harder to push a boat through heavier salt water, the boat floats higher, giving it a smaller cross section. The two effects cancel. Within the approximations described here, sailing on a lake of alcohol should be neither faster nor slower than sailing in ordinary water.

In general, maximizing S_0 (and boat speed) is accomplished only by sacrificing stability, safety, and sailing ease. An adventurous youth will enjoy the large S_0 of a light boat with large sails. Old salts are happier with the comfort of a boat with a modest downwind speed ratio.

2.4 Wind Shadow

Anyone trying to get out of the wind knows about wind shadows. Immediately downwind of any large object, the wind nearly vanishes, but as one moves away, the wind gradually returns to its original velocity. The same effect occurs with sailboats even though they are moving. A wind shadow extends downwind from every sail. Racing sailors should avoid one another's wind shadows. A sailor trying to catch boats ahead is not ashamed to cast his downwind shadow on competitors, thereby producing some curious downwind tactics, as suggested in Figure 2.6.

Figure 2.6   MC sailboat #2077 is in danger of losing wind because #2444 is close behind and in the wind path.

There is some confusing terminology relating to wind shadows. In fluid mechanics, a "wind shadow" is an example of a "wake." For sailors, wakes are the surface water waves produced by fast boats. They are described in Chapter 8.

Physical principles provide an estimate of the extent and severity of a wind shadow. As is always the case, the basic ideas come from Newton's laws. Newton tells us that forces always appear as equal and opposite pairs. In this case, the wind force driving the boat is exactly countered by the sail's force that slows the wind.

The apparent wind speed (not the true wind speed) in the wind shadow is decreased from V to $(V - \Delta V)$. The rate at which the wind recovers is described by $\Delta V(x)$, where x is the distance downwind of the shadowing sail. As one moves away from the sail and the wind recovers, $\Delta V(x)$ becomes smaller and smaller. For large enough x, $\Delta V(x)$ is insignificant. At the same time that the wind is recovering, the size of the wind shadow is growing. Its spreading width $L(x)$ and growing cross-sectional area $A(x)$ can also be estimated.

An approximation for the change in wind speed in the wind shadow starts with conservation of momentum. The sail decreases the wind's total momentum. However, after the wind passes the sail, there are no additional forces on the wind. This means the decrease in total momentum should not change as one moves downwind of the shadowing boat. This momentum change is the product $\Delta V(x) \cdot A(x)$. Just behind the sail where $x = 0$, the shadow area is roughly the sail area, $A(0) \cong A(sail)$ and $\Delta V(0) \cong V$, because there is essentially no apparent wind right behind the sail. Conservation of momentum then means that for all x

$$\Delta V(x) \cdot A(x) \approx V \cdot A(sail) \qquad (2.16)$$

The wind shadow grows both horizontally and vertically at about the same rate. Since $A(sail) \approx (L(sail))^2$ and $A(x) \approx L(x)^2$, where $L(sail)$ and $L(x)$ are the linear dimensions of the sail and the wind shadow, Equation 2.15 means

$$\Delta V(x) \approx V \left(\frac{L(sail)}{L(x)} \right)^2 \qquad (2.17)$$

As an example, Equation 2.17 means that by the time the wind shadow width has increased by a factor of 5, the decrease in wind speed will be only about 1/25 the wind speed. This 4% change may be difficult to notice.

The wind shadow is actually more complicated than just a decreased average wind speed. The edges of the wind shadow are vaguely defined. As one would expect, the decrease in wind speed, ΔV, varies smoothly and becomes largest at the center of the wind shadow. By interrupting the steady flow, the sail increases the random swirling of wind in the wind shadow. Thus, a sailboat sitting in the wind shadow of another boat is subject to both a decreased average wind speed and increased turbulent fluctuations, known by sailors as "dirty air." The rapid wind fluctuations in dirty air make sailing much more difficult.

Sailboats are not the only source of a wind shadow. A hill on a windward shore means a decreased wind velocity and increased turbulence. Since hills are bigger than sails, their wind shadows extend much farther. It is generally a good idea to avoid sailing directly downwind of steep hills.

How far does a wind shadow extend? That is the important question, but it is also the hardest question. An estimate of a wind shadow's vitality at distance x downwind of the shadowing boat uses some pretty "shadowy" reasoning. The result is based on ideas about turbulence originated by Ludwig Prandtl and others in the first half of the 20th century.

The turbulence caused by the shadow includes sideways winds, which allow the shadow's size to grow as it moves downwind. Assume the swirling speeds are comparable to the decrease in the mean speed ΔV. Then, in a short time δt, a sideways speed ΔV would cause the wind shadow to grow by a distance $\delta L(shadow) \approx \delta t \cdot \Delta V$. Here "$\delta$" stands for "very small change in." In the same short time δt, the wind shadow has moved downwind a distance $\delta x = \delta t \cdot V$. This means

$$\frac{\delta L(x)}{\delta x} = \frac{\delta t \cdot \Delta V}{\delta t \cdot V} = \frac{\Delta V}{V} \qquad (2.18)$$

Then using the relation $(\Delta V/V) \approx (L(sail)/L(x))^2$ from Equation 2.17 gives an estimate that requires either calculus or a willingness to cancel the δ's in $\delta L(x)/\delta x$. Either way, the result is

$$\frac{L(x)}{L(sail)} \approx \left(\frac{x}{L(sail)}\right)^{1/3} \tag{2.19}$$

Combining Equations 2.17 and 2.19, the decrease in the average wind speed in the shadow for $x > L(sail)$ is

$$\Delta V(x) \approx V\left(\frac{L(sail)}{x}\right)^{2/3} \tag{2.20}$$

Wind shadow physics is essentially described by Equations 2.19 and 2.20. These equations mean that the shadow size should be (very roughly) double the boom length at 8 boom lengths downwind. Its size will be tripled at 27 boom lengths downwind. At 8 boom lengths downwind, the shadowed wind speed has recovered to roughly 3/4 V. At 27 boom lengths, it is about 8/9 V. Although this makes qualitative sense, the numerical values are only rough guides. This wind shadow shape and the wind recovery curve are sketched in Figure 2.7.

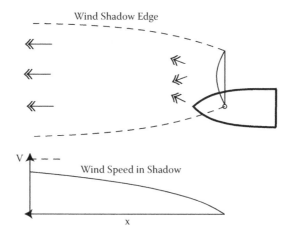

Figure 2.7 A rough outline of a wind shadow that grows slowly as the cube root of the downwind distance. The arrows in the wind shadow suggest the decreased wind speed and increased turbulence near the sail. Also shown is the recovery of the apparent wind speed as a function of the distance x. The loss of wind speed decreases as the 2/3 power of the distance downwind.

This theory has some pretty bold approximations, and the theoretical results do not exactly agree with sailor's experience. For example, sailing lore says wind shadows extend farther and are more of a nuisance in lighter winds. There is nothing in the theory described here (or my understanding) that explains a more prominent light wind shadow.

Wind shadows also occur for upwind sailing. The geometry and the details are different, and the upwind wind shadow extends in the direction of the apparent wind, which is described in Chapter 3.

The wind shadow is not the only consequence of the tit-for-tat symmetry of Newtonian mechanics. As the boat plows through the water, it drags water behind it and a small current follows the boat. A sailboat is a mechanism for transferring the motion (momentum) of the wind into motion (momentum) of the water. The wind pushes the boat which, in turn, pushes the water. The 800-fold difference between water and air densities means the water current following the boat is much less noticeable than the wind shadow in front of the sail. However, there is still a slight advantage in following directly behind another sailboat because of the current it drags behind. The opposite is true behind a power boat, because the propeller pushes the water backward.

2.5 Acceleration

Neither the wind speed nor the sailboat speed remains constant. Increases in the wind will accelerate the boat, but it takes some time for the boat to respond. Using Newton's laws ($F(total) = ma$) means

$$m^* \frac{d}{dt} u(t) = F_D(wind) - F_D(water) \qquad (2.21)$$

Here, $u(t)$ is the boat speed at time t, and the derivative $du(t)/dt$ is the acceleration or the rate at which the speed changes. The two terms on the right are subtracted because the force from the wind opposes the force from the water. The mass in $F = ma$ has been replaced by an "effective mass" m^*. Objects moving through the water drag some water with them, and thereby increase their effective mass. A famous but complicated calculation for a deeply submerged object moving through a fluid with no viscosity shows that the effective mass is increased by 50%, so $m^* = 3m(boat)/2$. This is another result that deserves a skeptical reception. Sailors certainly hope their boats never become deeply submerged objects, and the assumption of zero viscosity is surely wrong.

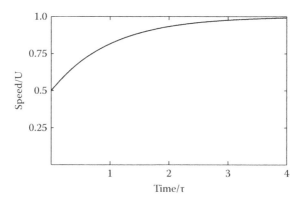

Figure 2.8   The time dependence of boat speed that is initially moving at half the steady-state speed U.

The equation for downwind acceleration simplifies when $S_0 = 1$, which is a typical value for the downwind speed factor. In this case, Equation 2.21 simplifies to

$$\frac{d}{dt}(u(t) - U) = -\frac{1}{\tau}(u(t) - U) \qquad (2.22)$$

Equation 2.22 is a precise way of stating the obvious. The boat slows down when $u(t)$ is greater than the steady-state speed U, and it will speed up if $u(t) < U$. An example result for the acceleration of a boat whose initial speed was only half of U is shown in Figure 2.8

The time it takes to approach the steady-state speed is the "time constant" τ. Nearly 2/3 of the speed change is accomplished in one time constant τ. The significance of this quantitative description is a relatively simple expression for the time constant, which is presented in the following section.

2.6 Examples

2.6.1 Force and Power

Force and power vary a great deal with the wind speed. The sail force is essentially given by Equation 2.7, written more explicitly as

$$F_D(wind) = \frac{C_D(sail)}{2}\rho(air) \cdot A(sail) \cdot V^2 \qquad (2.23)$$

In order to sail well in light winds, the sail area $A(sail)$ should be quite large. But then $F_D(wind)$ is dangerous in heavy blows. If it is too windy, sailors can reduce the force by decreasing the sail area. To keep the force constant, a doubled apparent wind speed V would require reducing the sail area by a factor of four.

The force is also proportional to the air density $\rho(air)$, which is roughly 10% smaller on a hot day than it is on a cold day. Altitude and barometric pressure also make a difference in the density. If a sailor takes a boat from sea level to a regatta near Denver, Colorado, the density $\rho(air)$ and the corresponding force of the wind will be decreased by about 15%. These variations in density (and the wind's force) could influence a sailor's decision on how much crew weight is needed or which sail to use. Even though high humidity makes muggy air feel heavy, water molecules are lighter than the oxygen and nitrogen molecules of dry air, so higher humidity means a smaller force. In practice, high humidity has a much larger effect on a sailor's mood than it does on the air density or sailing physics.

One can check out the force formula for a "typical" condition. An example is a Thistle sailboat shown in Figure 2.9 whose sail area is $A = 17.75$ m² (ignoring spinnaker) sailing downwind in a Fresh Breeze $W = 10$ m/s. For simplicity, assume the steady-state speed U is half the true wind speed W so $V = W/2$. Assuming the sail's drag coefficient is $C(sail) = 4/3$ means $F_D(Thistle : Fresh\ Breeze) \cong 370$ N. This is about half the force needed to lift a person off the ground. (For people living in the United States, Myanmar, and Liberia, a Newton is between a fifth and a quarter of a pound. For a traditional Britisher, one stone is 31 N.) Since the force is proportional to the square of the wind speed, the force in a Gentle Breeze, $W = 5$ m/s, is four times smaller, or about 92 N.

Physically, the force on the sail results from a pressure difference Δp between the windward and leeward sides of the sail. The force is the average pressure difference multiplied by the sail area A. For the downwind Thistle sailboat example, this gives $\Delta p(Fresh\ Breeze) \cong 21$ N/m². This pressure difference between the two sides of the sail is only one part in 5,000 of the 100,000 N/m atmospheric pressure. For all practical purposes, there is just as much air on the back side of a sail as there is on the front. In a Gentle Breeze with half the wind speed, the pressure difference is four times smaller.

Figure 2.9   The Thistle sailboat sailing approximately downwind. The raised spinnaker and lowered jib change the sail area estimate.

Wind powers the sailboat in a real sense because force times velocity is power (at least for downwind sailing). Multiplying a Fresh Breeze force on the Thistle by the boat speed of 5 m/s gives an estimated Thistle power as *Power*(*Thistle* : *Fresh Breeze*) ≅ 1,850 W. The 1,850 W is about the same as 2.5 hp, but the wind will push the boat faster than a 2.5 hp engine because the 2.5 hp rating is the maximum power in the drive shaft. The conversion of this power to force on the boat through the action of a propeller is not very efficient.

The force and power estimates can be performed for the boat of your choice. The force on a Laser sailboat is less than half that for a Thistle. On the other hand, the force on an America's Cup boat with a sail area of around 300 m² (again ignoring spinnaker) is F_D(*America's*) ≅ 6,250 N, which is about 42 hp. However, as is described in Chapter 10, it is never advantageous for an America's Cup boat to sail directly downwind.

For boats of any size, the force is proportional to the square of the speed, and the power is force times speed. In other words, the power is proportional to the cube of the wind speed. The power in Gentle Breeze (with half the wind speed) is only one-eighth the power in Fresh Breeze.

A sailboat provides a vivid illustration of the wind's power. One could attach a sailboat to an electric generator and harness the wind, but there is a more practical way to help supply the world with energy. A modern wind turbine (windmill) whose blades sweep out an area A does a better job in harvesting the power of the wind. For both sailboat and wind turbine, the power is proportional to the cube of the wind speed, which explains why wind turbines are of almost no use in locations where the average wind speed is small. In practice, wind turbines are designed with a maximum wind speed limit, so the *Power* $\propto W^3$ relation has an upper limit.

Wind turbine efficiencies are often compared to a theoretical maximum power called "Betz's law," which says,

$$P(\text{max; } Betz) = \frac{16}{27}\rho(air)W^3 A \qquad (2.24)$$

A comparison with sailboat power yields a similar expression. The sailboat power is

$$P = \frac{1}{2}C_D(sail)\rho(air)U(W-U)^2 A(sail) \qquad (2.25)$$

This power takes its maximum value when $U = W/3$. The wind does more work on a relatively slow sailboat because it must push harder. At this slow speed

$$P(Sail) = \frac{2}{27}C_D(sail)\rho(air)W^3 A(sail) \qquad (2.26)$$

The value of $P(sail)$ depends on the drag coefficient, $C_D(sail)$. Assuming this drag coefficient is roughly unity and assuming wind turbines are half as efficient as the Betz's law maximum, one concludes that sailboats are only one-fourth as effective as wind turbines in extracting the wind's energy.

Betz's law is not a rigorous bound because there are special situations where the drag coefficient C_D can become very large. In principle only, $P(sail) > P(\text{max; } Betz)$ is a possibility. This violation can occur because a very small object in the path of a very slow wind acquires a very large drag coefficient. This low Reynolds number limit is never of interest to sailors or the manufacturers of wind turbines. However, the low-speed and small-size limit is an important footnote in the history of physics. In 1909, Robert Millikan used the drag force on slowly falling tiny drops of oil to obtain the first measurement of the electron's charge.

2.6.2 Real Boat Speeds

The numerical estimates of force and power assumed the steady-state downwind sailboat speed, U, is half the wind speed, W. In principle, this can be checked using the previously derived formulas

$$U = \frac{S_0}{1 + S_0} W \qquad (2.10)$$

and

$$S_0^2 \cong \frac{C_D(sail)}{C_D(hull)} \frac{\rho(air) \cdot A(sail) \cdot L}{2m(boat)} \qquad (2.15)$$

Using the Thistle sailboat as a generic example, the mass $m(boat)$ of the boat and crew is typically 400 kg. The boat length L is 5.18 m, and the sail area is 17.75 m². This means

$$S_0^2 \approx \frac{1}{7} \frac{C_D(sail)}{C_D(hull)} \qquad (2.27)$$

The guess for the sail was $C(sail) = 4/3$. For a sphere, $C_D \cong 2/5$ (or less). If the hull had one-third the drag coefficient of a sphere, one obtains $S_0 = 1$, and thus $U = W/2$. It would appear that a streamlined design could make $C_D(hall)$ even smaller, which would imply sailboat speed significantly larger than half the wind speed. An ideal $C_D(hall)$ is difficult to achieve because of wake generation, stability considerations, and other practical limitations of hull shapes.

Since the square of the speed ratio is proportional to the sail area divided by the boat's mass ($S_0^2 \propto A(sail)/m(boat)$), the simplest way a sailor can increase downwind boat speed is to increase the sail area $A(sail)$ or decrease the boat mass $m(boat)$. This outcome can be achieved by raising a spinnaker or taking on a lighter crew. Attempting to go faster is not as rewarding as one would hope. For a "typical" case where $S_0 \cong 1$ and the boat speed is about half the wind speed, a 4% decrease of m or increase of $A(sail)$ results in a 2% increase in S and only a 1% change in boat speed. If one wishes to travel at three-frouths the wind speed downwind, one would have to achieve an unrealistic $S_0 = 3$. Although variations in S_0 are small, they are not insignificant for racing sailors.

On light boats, a change in crew weight can make a difference. The mass of a Laser sailboat is only 57 kg. In theory, a light Laser sailor whose mass is 57 kg will be faster downwind than an 87 kg Laser sailor. The theoretical speed difference is about 5%. At the end of a run that is 1 km long, the lighter sailor should gain about 50 m on the heavier sailor. The situation on lighter boats is actually more complicated because proper crew placement, which may be done more effectively by a heavier crew, can decrease the hull drag. Also, when it is windy the extra weight can increase stability even for downwind sailing. For heavier boats, the advantage of a lighter crew is much less significant.

The speed advantage of raising a spinnaker is more apparent. If the extra sail increases the effective sail area by 50%, then the boat speed will increase by about 10% (again, assuming $U/W \cong 1/2$). It is not so easy to estimate the effective area of a spinnaker because sails overlap.

2.6.3 A Check

The simplified theory says that on calm or windy days, sailboats should sail downwind at about half the wind speed. In practice, sailboats run into problems when the wind increases. Some typical results in Figure 2.9 show the ratio of downwind boat speed to wind speed for four sailboats. In order from slowest to fastest, they are the Cal 40, Beneteau 17.7, Grand Soliel 40, and the Fastcat 435. Except for the catamaran (Fastcat 435) these boats can achieve half the wind speed only in relatively light winds. Wake formation and overpowered sails at higher speeds are probably the major causes of decreasing U/W when the wind blows hard. A fairly large variation in S_0 is needed to produce the modest differences in boat speeds shown in Figure 2.10. The catamaran's 57% of the wind speed corresponds to $S_0 \cong 4/3$, which

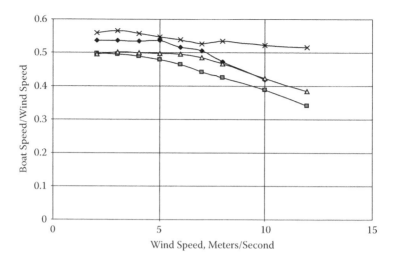

Figure 2.10 The ratio of downwind sailboat speed to wind speed for (from slow to fast) Cal40, Beneteau 17.7-shallow draft, Grand Soliel 40, Fastcat 435.

is significantly larger than the $S_0 \cong 1$ for the two slower sailboats. The wide stance of the catamaran allows it to have a relatively large sail area while maintaining a light weight, leading to a larger value of S_0.

Results shown in Figure 2.10 should be taken as qualitative indications rather than precise measurements. The wind is never steady and wave conditions are never the same. Figure 2.10 is also misleading because it suggests that there is little speed difference between sailboats. This is only true for downwind sailing. For other sailing directions, the speed difference is amplified, and fast sailboats can have twice the speed of slower boats.

2.6.4 Better Speed Calculations

The formula $U = WS_0/(1 + S_0)$ from Equations 2.10 is primitive. The simplest approximation for the speed ratio S_0 depends only on the mass m, sail area $A(sail)$, length L, and guesses for the drag coefficients. Although m, $A(sail)$ and L are basic to determining speed, many other sailboat properties enter into any realistic estimate of speed.

An alternative and practical way to find a formula for boat speeds is to compare the speeds of real sailboats in real sailing situations. An example formula that has been used for larger boats with keels is the

Schell Regression Formula

$$U = \frac{1}{K_1 - K_2\left(\frac{A}{m^{1/3}}\right) + K_3 A^2 - K_4\left(\frac{P}{J+E}\right) - K_5\sqrt{L} + K_6\left(\frac{d^2}{A}\right)} \qquad (2.28)$$

In this monster formula, $K_1, K_2, K_3, K_4, K_5,$ K_6 are numbers picked to give a good fit to observed speeds of different sailboats. As before, A stands for sail area, m is the mass of the boat, and L is its length. Also (and approximately), P and E are the height and length of the mainsail, J is the distance from the mast to the bow, and d is the depth of the keel.

The Schell Regression Formula doesn't look anything like Equation 2.10, $U = WS_0/(1 + S_0)$. As is usually the case, a phenomenological formula fits the data better than an oversimplified theory. However, there is no fundamental justification for Equation 2.28.

If one desires an accurate and realistic calculations of a sailboat speed, both the simple expression $U = WS_0/(1 + S_0)$ for downwind sailing and the Schell Regression Formula are far from the last word. Simple formulas can describe only simplified physics. Vastly more complicated calculations are needed to do a good job. Many equations relating many variables are needed, and some of these equations must be based on experimental results rather than theory. Computers are good at dealing with this messy business. Commercially available computer-generated velocity prediction programs give fairly accurate results, especially for standard heavy-duty sailboats. The most sophisticated calculations are closely guarded secrets, for obvious reasons.

2.6.5 Acceleration

The acceleration described by Equation 2.22 has a simple interpretation, as is illustrated in an example. Two identical sailboats are initially next to each other. They each wish to sail downwind. However, one of these boats is stopped dead in the water while the other is sailing with the steady speed U. It takes the stalled boat roughly one time constant τ to get moving. By the time the stalled boat is up to speed, the total distance lost is

$$X = U\tau \qquad (2.29)$$

Doing the algebra needed to derive the time constant gives a relatively simple expression for the distance lost.

$$X \approx \frac{3}{8} \frac{L}{C_D(hull)} \qquad (2.30)$$

Hull-drag coefficients are typically considerably less than unity. Roughly speaking, this means that when a sailboat completely stops, it will lose a couple of boat lengths (or more) before it is up to speed again. Since X does not depend on the wind speed, the distance lost is (approximately) the same in light air and heavy winds. In very light air, it takes a long time to regain speed. The lesson from all of this is "don't rock the boat," especially in light winds.

Although it was derived for downwind sailing, the distance X is a rough guide to the response of a sailboat for all sailing directions. An example is tacking when sailing upwind. Typically, one boat length is lost when one makes the roughly 90° turn needed to change the wind from one side of the boat to the other. Much of this lost distance occurs after the tack is finished and the boat is regaining speed. The lesson: the best tack is not the quickest tack; it is the tack in which the least speed is lost during the maneuver. A tactical consequence of this result tells a sailor that it is better to tack in a strong wind than a calm. The distance lost in each case is the same, but more time is lost in light air. In the final analysis, time is what counts.

This example is, like all the examples, an oversimplification. There is a second delay time that has been ignored. When a boat changes orientation or sail trim, it takes a little time for the wind pattern to settle down to its steady-state motion. So the sailboat that is stalled and suddenly aims downwind must wait for the pressure to develop on the sail. Generally, the time to accelerate the boat is the larger time, and loss of time should not be blamed on a lazy wind.

2.7 The Speed Limit

Downwind sailing has a speed limit. No matter how big the sail, and no matter how light the sailboat, sailing downwind faster than the wind is impossible. Only a sailboat that weighs nothing (a balloon) can keep up with the wind.

There is also a speed limit in special relativity. Only a zero mass particle (a photon) can move with the speed of light. But this is the end of the analogy. Light speed is associated with physics being the *same* in every coordinate system. Boat speed is limited because the physics is *different* in every coordinate system. In particular, a sailboat moving at wind speed is in a special coordinate system in which one feels no wind at all, so there can be no force on the sails.

The formulas presented here are approximations. Errors can be considerable. But the speed limit is absolute and does not depend on formulas. The explanation of this speed limit is the essence of downwind sailing.

- The boat is pushed ahead by the wind.
- The water pulls the boat back.
- The boat accelerates until the wind and water forces are equal.
- The boat's speed subtracts from the wind speed, so the wind force vanishes as the boat speed approaches the wind speed.

There is no speed limit for upwind sailing. Upwind sailing can be more exciting.

3

UPWIND—THE HARD DIRECTION

3.1 Overview

Sailing against the wind is mysterious. It is easier to sail upwind than to understand it. Sitting on a sailboat, you see the water moving past from bow to stern. The wind is coming slightly from the side, but it can also be mostly from the bow. How can a sailboat move against both the wind and the water?

Lift is the key. Sailing upwind, as shown in Figure 3.1, is possible only because the wind's force is not parallel to the wind direction, and the water's force is not parallel to its motion relative to the boat. Combining lift from the wind with lift from the water produces the miracle of upwind sailing. Lift also produces the miracle of flight, but sailing upwind is more complicated. Birds and airplanes must contend with only a single fluid—air. Sailing is the story of two fluids (air and water), and the boat that lies at their interface.

3.1.1 Lift and Drag

For sailors, lift and drag are the essence of fluid mechanics. Drag is simple. When wind or water pushes something in the direction it is moving, it is a drag. The drag force is denoted F_D in general. When the distinction is needed, drag from the wind and water are denoted $F_D(wind)$ and $F_D(water)$. Lift is drag's exciting partner, which pushes perpendicular to the drag. For an immediate experience of lift and drag, stick your hand out the window of a speeding car and feel the variations of the force as you tilt the orientation of your hand. Lift lifts your hand and drag pulls it back. The lift force is denoted F_L, $F_L(wind)$ or $F_L(water)$, as is appropriate.

The sum of the perpendicular lift and drag forces from either the wind or the water is a vector, denoted $\vec{F}$, $\vec{F}$ (wind) or $\vec{F}$ (water). Little arrows

Figure 3.1   Thistle sailboats sailing upwind. The boat on the right is on starboard tack because the wind is coming from its starboard side (right side when facing forward). The other boat is on port tack. (Photograph by Sally Snowden. With permission.)

over a total force mean it has both magnitude and direction. Examples of lift and drag forces and their vector sums are shown in Figure 3.2.

3.1.2 Wind Direction

Which way is the wind blowing? This is a tricky question on a sailboat. The air is moving (wind), the boat is moving, and sometimes the water is moving (current). The boat's velocity with respect to the water is $\vec{U}$.

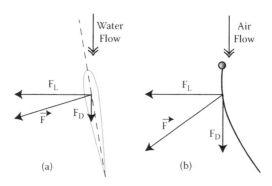

Figure 3.2   Lift and drag forces for shapes similar to the cross sections of a centerboard (a) and a sail (b).

The arrow over the $\vec{U}$ means it is another vector with both magnitude and direction. Without the arrow on top, U means the speed without reference to a direction. The "true wind" velocity (with respect to the water) is another vector $\vec{W}$. The apparent wind velocity $\vec{V}$ represents the wind observed on the moving sailboat. *Note*: Traditional terminology for wind directions ($\vec{W}$ and $\vec{V}$) is confusing. The velocity vector for a "north wind" points south, which is the direction the air is moving.

The true wind velocity $\vec{W}$ is the vector sum of the boat's velocity $\vec{U}$ and the apparent wind velocity $\vec{V}$:

$$\vec{W} = \vec{V} + \vec{U} \qquad\qquad (3.1)$$

A special case of Equation 3.1 for downwind sailing is Equation 2.8 where (in this special case) $V = W - U$ means the apparent wind speed is obtained by subtracting the boat speed from the true wind speed. For sailing in any other direction, the vectors must be treated with respect. Graphically, Equation 3.1 describes a "velocity triangle," with sides of length W, U, and V. An example is shown in Figure 3.3. The three angles of the velocity triangle are defined in the sense used by sailors, which is a little unconventional. The "true wind angle," w, is the

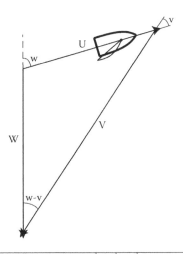

Figure 3.3   A velocity triangle illustrating $\vec{W} = \vec{V} + \vec{U}$. The smaller letters w and v are the angles between the wind source and the boat velocity, and ($w - v$) is the shift in the apparent wind direction.

angle between the direction the true wind is coming from and the direction the sailboat is moving. A boat pointing directly at the wind corresponds to $w = 0$, and downwind is $w = 180°$. The "apparent wind angle" v is defined in the same eccentric way, with the apparent wind replacing the true wind. The third angle is the "apparent wind shift," $(w - v)$. It is the angle the apparent wind is rotated from the true wind direction by the boat's motion.

3.1.3 Forces

Upwind sailing is a vector generalization of the downwind case. The steady-state condition means the wind and water forces are equal and opposite, so

$$\vec{F}(wind) + \vec{F}(water) = 0 \qquad (3.2)$$

On the sailboat, the wind that matters is the apparent wind. So the forces on a sailboat are expressed most directly in terms of the apparent wind speed V and the apparent wind angle v.

For the example in Figure 3.4, the apparent wind $\vec{V}$ is a "north wind" blowing south, so the wind's drag is directed south (bottom of page), and the wind's lift is to the east (right). The water's drag is opposite the boat's velocity $\vec{U}$, and the lift (perpendicular to $\vec{U}$) is produced by a small sideways motion of the sailboat. In this example, the angle v between $\vec{U}$ and $(-\vec{V})$ is 50°, and the boat's orientation is rotated an almost imperceptible 4° from $\vec{U}$ toward $(-\vec{V})$. Because the boat is moving, the true wind $\vec{W}$ in this example is from a direction west of north.

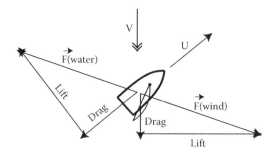

Figure 3.4   A force diagram showing the lift and drag components of the wind and water forces on a sailboat. When sailing at a constant speed, $\vec{F}(wind)$ and $\vec{F}(water)$ are equal and opposite.

At this stage, one could look at the force diagram and say "OK, that's how it works. Let's go sailing." But if sailing does not go as planned, one may be motivated to ask how the forces could be changed to make sailing faster, safer, or more comfortable. In principle, sailing performance can be improved through a better understanding of what pushes the sailboat. In practice, my sailing fiascos provide ample evidence that sailing physics alone does not produce a winning sailor.

Sailboat forces can be pretty confusing. Iceboating may not be easier than sailing, but it is easier to understand. So this example is considered first.

3.2 Iceboats

Until the 20th century, iceboating was the fastest known mode of human transportation. In 1888, the first automobile speed record was established at 17.5 m/s. For trains, the record speed in 1890 was 40 m/s. At essentially the same time, 1888, an iceboat on the Hudson River was reported to travel faster than the car or the train at nearly 48 m/s. An article from the January 22, 1871, edition of *The New York Times* describes a race between a "lightning train" and two iceboats, the *Zephyr* and the *Icicle,* on the Hudson River near Poughkeepsie, New York. The iceboats won. If the same race were held today, an iceboat could still win because trains traveling along the Hudson don't go very fast. Unfortunately, the Hudson at Poughkeepsie hasn't frozen solid for many years. Trains haven't changed much in over a century, but they don't make winters like they used to. Modern iceboats, such as those in Figure 3.5, are generally much smaller than the iceboats of a century ago.

The extraordinary speed of iceboats seems to defy logic. They can sail at four (some claim five) times the speed of the wind. When the iceboat speed is greater than the true wind speed, the apparent wind is unavoidably from the bow. Only the magic of lift forces allows such speeds.

Friction destroys tidy physics theories. This is certainly the case for sailing in water. The water's drag force means one must construct messy diagrams and resort to computer programs to accurately estimate speed. Even the bare-bones description of sailboat speeds presented in Section 3.3 is not really simple. For iceboating, the ice friction is so small that an idealized frictionless model provides a reasonable first approximation of this fast but cold version of sailing.

Figure 3.5   No matter which direction they are sailing, iceboat sails should be trimmed in tight. (Photograph by Stéphane Caron. With permission.)

3.2.1 Iceboat Forces

Assume the idealized runners on an iceboat present no drag force for forward motion, even though they completely eliminate sideslip. If an iceboat equipped with these perfect runners were to zip along on a frictionless ice surface, one might think it could be accelerated without limit. That is not the case because the apparent wind direction shifts toward the bow as speed increases. When the iceboat goes fast enough, the apparent wind blows it backward. The idealized iceboat will accelerate until the wind's accelerating force vanishes. As shown in Figure 3.6, the approximation of vanishing ice drag simplifies the force diagram. There are only three nonzero force components.

Because the ice force $\vec{F}(ice)$ is entirely lift, it is perpendicular to the iceboat's velocity. The opposing wind force $\vec{F}(wind)$ is also abeam. For this geometry, the apparent wind angle v is also the angle between the wind's lift and the wind's total force, as is shown in the Figure 3.6. Trigonometry (tan = *opposite/adjacent*) relates the apparent wind angle

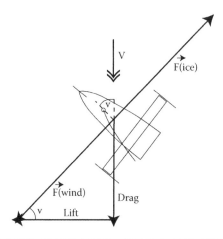

Figure 3.6 In the approximation of vanishing ice drag, the ice force $\vec{F}(ice)$ and the wind force $\vec{F}(wind)$ are both perpendicular to the iceboat velocity.

to the lift-and-drag forces. Omitting the unnecessary (*wind*) qualifier,

$$\tan(v) = \frac{1}{F_L/F_D} \rightarrow \frac{1}{[L/D]} \qquad (3.3)$$

In practice, an iceboat accelerates until the apparent wind angle is as small as possible. This occurs when the lift-to-drag ratio is as large as possible, and the maximum ratio is denoted $[L/D] = \max\{F_L/F_D\}$. This result is peculiar. It says the apparent wind angle is always the same because $\tan(v) = 1/[L/D]$, and $[L/D]$ is a property of the iceboat, not the direction or magnitude of the wind. Iceboats (and their sails) are designed to maximize $[L/D]$, so the apparent wind angle v is as small as possible. As a result, the wind is always blowing in your face. Although chilled iceboaters might wish it were different, the wind is never at your back.

3.2.2 Iceboat Speed Diagram

When sailing in a sailboat or an iceboat, what direction should one sail to go most quickly upwind or downwind? A speed diagram (also called polar diagram, or just "polar") explains everything and tells a sailor which direction to sail. Directions on a speed diagram are like directions on a map. North, corresponding to the true wind angle $w = 0$, is at the top. East with $w = 90°$ is to the right and so on. A point at the center of the map (the origin) is the initial position of a sailboat or iceboat. A curve surrounds this central point. Each point on the curve corresponds to a boat's position after sailing in the given direction for a fixed time (say, 1 min). For each direction, the distance to the curve is proportional to the boat's speed, so the speed diagram is really a graphical representation of speed. Since neither iceboat nor sailboat can sail north in a north wind, the speed diagram shows a range of directions close to north where sailing is impossible.

The speed diagram for the idealized iceboat has a surprisingly simple shape. There are two keys to an algebraic construction of the iceboat speed diagram. The first key is Equation 3.3, $\tan(v) = 1/[L/D]$, which was derived by ignoring the ice drag. The second key is a "speed-angle formula," which can be derived from the velocity triangle shown again in Figure 3.7. This time the little sailboat has been omitted and an extra line of length d has been added to the triangle. The extra line is perpendicular to the side V and meets the sides U and W at their vertex. Trigonometry $(\sin(\theta) = opposite/hypoteneus)$ means $d = U\sin(v) = W\sin$

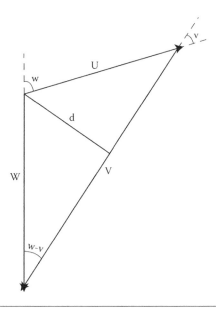

Figure 3.7   The extra line labeled d in the velocity triangle allows one to derive the iceboat speed equations that give the speed, U, as a function of the true wind angle, w.

obtains the speed-angle-formula

$$U(w) = W \frac{\sin(w - v)}{\sin(v)} \qquad (3.4)$$

nged a little. Since the iceboat speed depends on
:, w: $U \to U(w)$.
;ufficient to show that the idealized iceboat speed
ible-circle of Figure 3.8.

f Iceboat Speed Diagram

Algebra and Equations 3.3 and 3.4 yield a simple double-circle form for the iceboat speed diagram.

The demonstration comes in three parts: Part 1 presents the equation for a circle in polar coordinates; Part 2 shows the speed-angle formula is really the circle equation of Part 1; Part 3 adds the second circle.

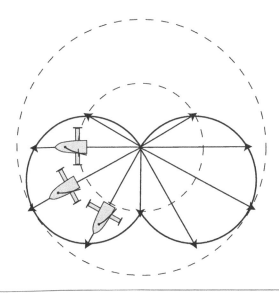

Figure 3.8   The speed diagram for an idealized iceboat whose maximum speed is twice the wind speed. The dotted circles indicate the wind speed and double the wind speed.

Part 1: In polar coordinates, $r = \sin(\theta)$ describes a circle. To show this, multiply both sides of the equation by r to obtain $r^2 = x^2 + y^2 = r\sin(\theta) = x$, which can be rewritten $(x - 1/2)^2 + y^2 = 1/4$. This is the equation of a circle with unit diameter centered at $x = 1/2$, so the circle passes through the origin.

Part 2: Starting with the speed-angle formula, $U(w) = W\sin(w - v)/\sin(v)$. Let $r = U(w)\sin(v)/W$ and $\theta = (w - v)$. Since W and v are fixed, $U(w)\sin(v)/W$ describes a circle with unit diameter as θ (or w) is varied. Multiply by $W/\sin(v)$ to obtain a circle with diameter $W/\sin(v)$ when $U(w)$ is plotted as a function of w. Because θ has been replaced with $(w - v)$, the circle is rotated so that it passes through the origin when $w = v$.

Part 3: The circle of Part 2 describes sailing on a port tack. Changing the sign of w and v gives the mirror image circle for sailing on starboard. Combining the two circles produces the double-circle form of the idealized iceboat speed diagram shown in Figure 3.8.

3.2.4 Iceboat Speed Diagram Interpretation

The idealized iceboat speeds are proportional to the wind speed. Thus, only one speed diagram is needed, and distances from the origin of this diagram represent the ratio of the iceboat speed to the wind speed. The two dotted circles in Figure 3.8 denote speeds equal to and double the wind speed. Nine arrows have been drawn to illustrate the speed for some representative directions. Reading top to bottom, the arrows on the right correspond to port tack directions that produce (1) the optimum velocity for upwind sailing, (2) sailing perpendicular to the true wind, (3) the velocity with the largest speed, which is double the wind speed in this example, (4) the optimum velocity for downwind sailing, and (5) downwind sailing with a speed equal to the wind speed. The superposed little iceboats show that the sail angle is the same for every sailing direction because the apparent wind angle v never varies for this idealized iceboat model.

It is never fastest to sail an iceboat directly downwind. If you want to go south, jibing downwind (sailing southwest and then southeast) is quicker. The geometry of a circle means the iceboat's optimum upwind velocity (1) is perpendicular to its optimum downwind velocity (3). This is not the case for ordinary sailboats, and jibing downwind is not a smart strategy for slow sailboats. Of course, all sailing craft of any speed must tack upwind (sailing northwest and then northeast).

The speed diagram in Figure 3.8 was drawn for an iceboat of moderate speed in ideal conditions, where the maximum speed is double the wind speed. For this case, $\sin(v) = 1/2$, which means the apparent wind angle is $v = 30°$ and $\tan(v) = 1/\sqrt{3}$. Thus, the sail's lift-to-drag ratio is $[L/D] = \sqrt{3} \cong 1.7$, which is not a difficult lift-to-drag ratio to achieve.

In this idealized view, iceboats are distinguished only by their lift-to-drag ratios. A comparison to two iceboats with different values of $[L/D]$ is shown in Figure 3.9. The inner dotted curve is the same speed diagram as shown in Figure 3.8, corresponding to an iceboat whose maximum speed is twice the wind speed. The outer speed diagram corresponds to a very fast iceboat whose maximum speed is 3.5 times the wind speed. To achieve this speed requires $[L/D] = 3\sqrt{5}/2 = 3.35$. Sails can provide a ratio of lift to drag forces that is this large. However, the ratio $[L/D]$ includes the wind drag on the hull and the cold iceboater. Fast iceboats are designed to be as streamlined as possible to minimize this drag because it has a dramatic effect on the speed.

Figure 3.9 also shows that the faster iceboat makes quickest progress to windward at a smaller wind angle w than the slower iceboat. This is often not the case for sailboats.

Regardless of speed, these iceboat speed diagrams always have two points in common. One is the origin corresponding to zero speed

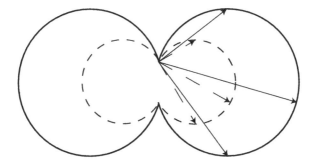

Figure 3.9 A comparison of two iceboat speed diagrams whose maximum speeds are twice and 3.5 times the speed of the wind. The arrows show the directions these two iceboats would sail for maximum progress to windward, for maximum speed, and for maximum progress downwind.

when sailing directly at the wind. The second common point is at the vertex due south of the origin that corresponds to inefficient downwind sailing at the speed of the wind.

3.2.5 Ice Friction

Even the smoothest ice slows an iceboat and destroys the simple elegance of the double-circle iceboat speed diagram shown in Figure 3.8. Reality shrinks and distorts the circles. When a smooth object with mass m moves over smooth ice, the drag force is approximately described in terms of a friction coefficient μ.

$$F_D(ice) = \mu mg \qquad (3.5)$$

The coefficient of friction for ice can be as small as 0.02, and it actually decreases a little with speed. This is completely different from the lift and drag forces of fluid mechanics that are proportional (approximately) to the square of a speed. The weight of a DN (named for the *Detroit News*) iceboat plus crew is roughly mg = 1300 N. Thus, the drag force computed from the coefficient could be as small as 25 N, which is the force needed to lift two champagne bottles. The sail area of the DN is only about 5.5 m². However, if the iceboat speed is really several times the wind speed, the corresponding sail force can be hundreds of Newtons. Thus, the friction force really can be a small fraction of the wind force on a windy day. When this is the case, the idealized double-circle speed diagram is a reasonable approximation.

Ice is seldom ideal. Roughness or a little snow significantly increases the coefficient of friction. Also, the iceboat runners must cut into the ice in order to prevent sideslip. This will also increase the drag, but the magnitude of the effect is difficult to estimate. The ability of iceboats to move very fast is proof that the drag is small.

3.3 Sailboat Speeds

A sailboat has a speed diagram analogous to the iceboat speed diagram, but calculations of sailboats speeds are long and tedious. One can avoid the tedium by skipping to the results in Section 3.3.7.

Sailboats are slower than iceboats because a sailboat must push the water out of its path. Water's drag complicates the forces and the geometry. Even computations based on numerical simulations and measured forces yield only approximate results. Despite the complexity, three key sailboat properties go a long way to determine sailboat speeds. These are:

1. The downwind speed ratio S_0. This is the same S_0 used to describe downwind sailing.
2. The wind's lift-to-drag ratio $[L/D](wind)$, evaluated at maximum lift.
3. The water's maximum lift-to-drag ratio $[L/D](water)$.

The following will describe sailboat speeds entirely in terms of these three numbers. If sailboats had feelings, they would be insulted at the prospect of being described by only three numbers, and they would rightly insist that they deserve a more sophisticated characterization.

AN ASIDE ON PHYSICS AND APPROXIMATIONS

Despite appearances to the contrary, one goal of physics is simplicity. A search for simplicity can lead to important physical discoveries, like the heliocentric theory of our solar system. But sailboats are not solar systems. There is nothing profound to be discovered through a search for sailboat simplicity. Ignoring details by reducing a sailboat to just three numbers can only introduce errors. However, including all the details can overwhelm our ability to understand.

When physics strips away complicated details, the approach is called a "model calculation," a "toy problem," or a "spherical cow approximation." Generating a sailboat's speed diagram in terms of just three numbers is the spherical cow approximation of sailing.

For many difficult problems, the model calculation is just the first step toward a correct and detailed understanding. This applies to the sailing model as well. One can examine the results, see their shortcomings, and add improvements so the spherical cow gains a little shape.

BACK TO THE SAILBOATS

Even the naïve approach described here yields results only after a sobering six-step program. These steps are:

1. *Lift and drag phenomenology*
   The standard notation describing lift and drag means the really hard problems of fluid mechanics can be buried in the lift and drag coefficients, C_L and C_D.
2. *Centerboard lift and drag*
   The lift of a streamlined object like a centerboard can be surprising large. The drag is surprisingly small. These combine to allow efficient

sailing. The theory behind lift and drag is complicated and an example shows how one may easily obtain erroneous results.

3. *Pushing the sailboat*

The force needed to push a sailboat is the first piece of the sailing puzzle. This force is complicated because both the lift and drag components depend on the sailboat's speed and its leeway (sideways motion).

4. *Wind's lift and drag*

The force produced by the wind is the second piece of the sailing puzzle. The wind's force is also complicated because sailboats must sail both upwind and downwind. Upwind sailing is "aerodynamic," and the wind skims by sails at a small angle of attack. Downwind "impact" sailing is more intuitive. The transition between aerodynamic sailing and impact sailing has special significance in the characterization of sailboat speeds.

5. *Wind and water forces*

The two-piece sailing puzzle is solved by requiring equality of the force needed to push the sailboat and the force produced by the wind. The equality is obtained by superposing graphical representations of the wind and water forces. The result is an expression for the sailboat speed U in terms of the apparent wind speed V and the apparent wind angle v.

6. *Sailboat speed diagrams*

Additional geometry is needed to describe the boat speed U in terms of the true wind speed W and true wind angle w.

3.3.1 Step 1: Lift and Drag Phenomenology

Lift and drag forces produced by a fluid moving past an object have some universal characteristics.

- Lift and drag are roughly proportional to the square of the fluid speed, U (for the water) or V (for the wind).
- They depend on the size, shape, and orientation of the object in complicated ways.
- They can both be changed by the surface texture in complicated ways.
- They are both influenced by fluid turbulence.

The standard notation that characterizes lift and drag forces (F_L and F_D) recalls Newton's impact theory (Section 2.2.1). Corrections to the impact theory for the drag are expressed in terms of the drag coefficient C_D. A lift coefficient C_L is defined analogously. Thus, the drag and lift forces are written as an extension of Equation 2.7.

$$F_D = \frac{C_D}{2} \rho \cdot A \cdot V^2 \qquad (3.6)$$

and

$$F_L = \frac{C_L}{2} \rho \cdot A \cdot V^2 \qquad (3.7)$$

Here A is an area and ρ is the fluid density. The apparent wind speed is V. The analogous expression for the water force replaces V with the boat speed U. Although the drag and lift coefficients C_D and C_L hide a multitude of complexities, they sometimes appear to be quite simple. For example, when sailing downwind, A is the sail area, and a typical drag coefficient for a flat sail is C_D(*downwind sail*) $\approx 4/3$. Curvature of the sail can increase the drag coefficient. There is no lift downwind, so C_L(*downwind sail*) = 0.

3.3.2 Step 2: Centerboard Lift and Drag

Reasonably efficient upwind sailing is only possible because of the centerboard. ("Centerboard" is a shorthand term that means either centerboard or keel.) Wind tries to push a sailboat sideways. The centerboard's role is to limit leeway (sideways motion) by supplying a large portion of the lift force, labeled F_L in Figure 3.10.

The shape shown in Figure 3.10 resembles the cross section of a typical centerboard. The similarity to the cross section of a bird's wing or the top view of a fast fish is not a coincidence. Evolution has done a remarkably good job of finding streamlined shapes without the aid of a computer. However, since "fish gotta swim, birds gotta fly," the analogy can be overstated.

The lift and drag forces depend on the direction of the fluid flow. When the centerboard is aligned with the north-to-south fluid flow, the "angle of attack" θ and the lift force vanish. The drag is also quite small because of the streamlined shape. Using A as the front cross sectional area, the drag coefficient can be as small as 1/30th of the drag coefficient of a flat plate facing the flow with the same area. For example, if a centerboard extends one meter into

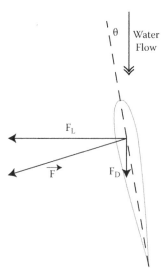

Figure 3.10   The lift and drag forces F_L and F_D depend on the angle of attack θ.

the water, is 3 cm wide, is moving at 5m/s in Fresh Breeze, and is characterized by a drag coefficient that is 20 times smaller than a flat plate,

$$F_D(centerboard : fresh\ breeze) \approx \frac{1}{2}\frac{4}{3} \times \frac{1}{20} \times 1000 \times \frac{3}{100} \times 5^2 \text{ N}$$

$$F_D(centerboard : fresh\ breeze) \approx 25 \text{ N} \qquad (3.8)$$

This estimate is small compared to that of the total drag estimate for a Thistle sailboat in Fresh Breeze (around 370 N). Leeway is not important for downwind sailing, so the extra 25 N of drag can be eliminated by raising the centerboard. The relations between the drag coefficient, the downwind speed factor and the boat speed (Equations 2.9 and 2.14) mean that raising the centerboard should increase the speed by not more than two percent.

   When the centerboard orientation is changed so the angle of attack is no longer zero, the lift grows rapidly with increasing θ. The drag also increases, but less spectacularly. It is a fortuitous peculiarity of fluid mechanics that the lift is linear in the attack angle but the drag increase is proportional to the square of the attack angle. For small angles, the drag and lift forces are accurately represented by two fairly simple equations.

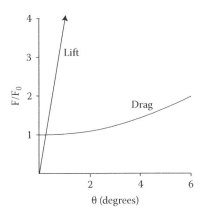

Figure 3.11   The dependence of lift and drag forces on the angle of attack. The forces are scaled by F_D, which is the drag for $\theta = 0$. At larger angles, the forces are more complicated.

$$F_D(centerboard) \cong F_0\left[1 + \left(\frac{\theta}{d_0}\right)^2\right] \qquad (3.9)$$

$$F_L(centerboard) \cong F_0\left(\frac{\theta}{l_0}\right) \qquad (3.10)$$

Here F_D is the centerboard drag when the angle of attack is zero. Two additional angles appear in these equations; d_0 is the angle at which the drag doubles and L_0 is the angle at which the lift equals the drag F_D. For the streamlined shape of a typical centerboard, the drag doubling angle d_0 is typically 5° to 8°. The linear increase of the lift force F_L is more dramatic and l_0 is typically only a small fraction of one degree. A sketch of the angular dependence of the lift and drag forces described by Equations 3.9 and 3.10 is shown in Figure 3.11.

The forces described by Equations 3.9 and 3.10 and shown in Figure 3.11 are valid only for small angles of attack. As θ becomes larger (typically more than 10°), more complex fluid motion often leads to abrupt increases in drag without any further increase in lift. The breakdown of the simple picture means lift is typically a maximum for θ not exceeding 20°. These are not large angles. The angle θ in Figure 3.9 is about 12 degrees, so the centerboard shown in Figure 3.10 would produce much more lift than drag.

These lift and drag characteristics apply as well to a sailboat's rudder. A sailboat is steered by the lift force on the rudder. Changing the rudder orientation changes the angle of attack θ and produces a lift that pushes the stern in the appropriate direction. If the rudder is turned sharply, the associated drag increase can be quite large, and this slows the boat. If the rudder is turned so sharply that the angle of attack is 20° or more, the lift can decrease. Since extreme steering is unproductive, sailors should treat their tillers (or wheels) gently. When a sailboat is making a turn, as from port to starboard tack, the water flow past the rudder is no longer from bow to stern. This means a larger, but gradual, increase in rudder angle with respect to the boat's axis is justified.

3.3.3 Where Is the Theory?

The relatively large lift illustrated in Figure 3.10 is a key to successful sailing because it allows one to design sailboats and sails with large lift-to-drag ratios. It is also one of many surprising results of fluid mechanics because a simple extension of Newton's impact theory fails to describe lift. If Newtonian common sense prevailed, sailing would be no fun at all.

Warning: the following attempt to describe lift using common sense is wrong.

Assume the impact theory remains valid for calculating lift, so molecules are little bullets bouncing off a surface. This is illustrated in Figure 3.12, where water molecules are reflected from a thin centerboard.

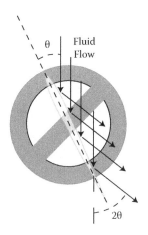

Figure 3.12   A wrong demonstration of lift based on the impact theory. Molecules bouncing elastically from a surface at an angle θ would have their direction changed by 2θ.

When the centerboard angle with respect to the fluid flow is θ, the flux of molecules that hit the side of the centerboard is proportional to $\sin(\theta)$. Elastic scattering means the molecules should be scattered by an angle 2θ. The horizontal momentum given to a reflected molecule is proportional to $\sin(2\theta)$. Equal and opposite forces mean the total force on the centerboard is proportional to both the particle flux and the momentum given the molecules. Thus

$$F_L(WRONG) \propto \sin(\theta)\sin(2\theta) \rightarrow \theta^2 \qquad (3.11)$$

Stated more qualitatively, it seems reasonable amount of fluid deflected should be proportional to the angle, and the deflection should also be proportional to the angle. Multiplying these gives the incorrect and relatively insignificant lift proportional to θ^2.

The following rationale for the linear dependence on the angle of attack is either profound or vacuous, depending on your viewpoint. The smooth nature of physical quantities makes one believe that lift can be written as a power series in θ, so the angular dependence of the lift force is

$$F_L(\theta) = a + b\theta + c\theta^2 + \dots. \qquad (3.12)$$

Since the lift vanishes when $\theta = 0$, the coefficient a must be zero. Also, the lift force changes sign when the angle of attack is reversed and $\theta \rightarrow -\theta$. Since θ^2 is always positive and does not change sign, the coefficient c must also vanish. If no higher terms in the series are considered, the only possibility is

$$F_L(\theta) = b\theta \qquad (3.13)$$

This is the result characterized by Equation 3.9 and shown in Figure 3.11. Unfortunately, a symmetry argument can not determine the coefficient b or the magnitude of the lift force. The steps leading to Equation 3.13 also fail to explain why the impact theory gives the wrong result.

It would be much more sensible at this point to simply provide the correct calculation of lift and drag. If these forces were nailed down, one could proceed to a solid quantitative description of sailing. Sadly, fluid mechanics does not provide easy answers. So for now, accept the numbers characterizing lift and drag as quantities determined by experiments or computer simulations. Fluids and the problems of calculating lift and drag are discussed in Chapter 8.

3.3.4 Step 3: Pushing the Sailboat

One can determine sailboat speed only if one knows how hard it is to push the boat through the water. Because sailboats make leeway (move sideways) the force needed to keep the sailboat moving depends on both its speed and its leeway. The centerboard will be considered first. Then a generalization gives the force needed to move the whole boat.

Assume the wind supplies the force needed to push a centerboard directly north. Moving north is equivalent to the water flowing south past the centerboard. If the centerboard is aligned north-south, the required force will be north because there is no lift. However, if the centerboard is aligned west of north as in Figure 3.10, the force to the north must be increased because of increased drag and the lift means the wind must also push to the east.

Signs are confusing. The force needed to move the centerboard is exactly opposite the force $\vec{F}$ shown in Figure 3.10, which is the water's force on the centerboard. Focusing on the force needed to push the centerboard rather than the forces shown in Figure 3.10 appears to introduce a confusing and annoying sign change. However, this annoying reversal simplifies the graphical representations that follow. The force needed to move the boat is the same (no sign change) as the force provided by the sails, so the two forces can be compared on a single graph.

For a given centerboard speed U, one can eliminate the angle of attack θ by combining $F_D = F_0(1 + \theta/d_0)^2)$ with $F_L = F_0(\theta/l_0)$ (the lift and drag forces of Equations 3.9 and 3.10). This yields

$$F_D = F_0 + \frac{1}{F_0}\left(\frac{l_0}{d_0}\right)^2 F_L^2 \qquad (3.14)$$

Thus, if the drag F_D is plotted as a function of the lift F_L, the curve has the shape of a parabola. This makes sense. If lift is required of a centerboard, its drag will increase. The shape of the parabola depends on the centerboard speed because F_0 is proportional to the square of the speed.

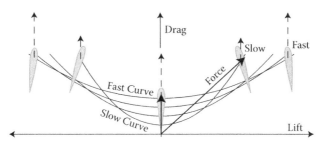

Figure 3.13 Each of the four curves shows the force needed to keep a centerboard moving north (toward the top of the page). The four curves correspond to pushing the centerboard at four different speeds.

Figure 3.13 shows four different parabolas obtained from Equation 3.14 with different boat speeds and thus different values of F_0. For each parabola, a vector from the origin to a point on the curve gives the magnitude and direction of the force needed to keep the centerboard moving north. Two example force vectors are shown in Figure 3.13. These are the darker arrows with large heads, one of which is labeled "Force." The centerboards decorating the parabolas illustrate the orientation associated with the force. Angles and scales in Figures 3.13 and 3.14 are exaggerated for the purposes of illustration.

The force needed to push the centerboard is only part of the total force needed to move the sailboat. The relatively large drag and limited lift of sailboat hulls means the centerboard produces most of the lift and the rest of the boat produces most of the drag. The extra drag changes the coefficients in Equation 3.14 and the details of Figure 3.13, resulting in the steeper parabolas shown in Figure 3.14.

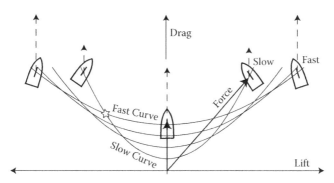

Figure 3.14 The drag and lift forces needed to keep a boat moving north. Each curve corresponds to a different speed. Even though the boats are not aiming north, the applied force pushes them north.

The small "star" placed in Figure 3.14 indicates a point where the "fast curve" and the "slow curve" intersect. This intersection means that two identical boats with the same sail force can both travel north, but at two different speeds. Whenever a boat appears to be moving slowly and sliding sideways, the boat is wallowing on a "slow curve" instead of zipping along on a "fast curve." Slow sailing typically results from aiming too close to the wind. The boat loses speed; less speed means less lift and more leeway; more leeway means the boat must aim even more toward the wind. Temporarily changing course away from the wind to gain speed allows a transition to the "fast curve" and a happier sailor.

Sails are not shown on the boats in Figure 3.14 because the water force is determined only by the orientation and speed of the boat. The next section considers wind force.

3.3.5 Step 4: Wind Lift and Drag

In some ways, the force generated by the wind $\vec{F}(wind)$ is analogous to $\vec{F}(water)$. There are lift and drag components whose sum is the total force vector. An idealized plot of $\vec{F}(wind)$ is shown in Figure 3.15. The horizontal lift axis is denoted "L" and the vertical drag axis is "D."

The graphical representation of $\vec{F}(water)$ has additional structure because the angle of attack between the sail and the wind is not always

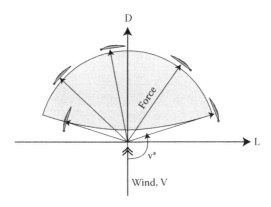

Figure 3.15   The wind's force vector $\vec{F}(wind)$ can be any point on the boundary of the shaded region. Example sail alignments and the corresponding forces are shown for five points on the curve. The apparent wind angle v^* to the "kink" in the curve has special significance.

small. The full range of angles generates a closed curve (with shaded interior) instead of a simple parabola. Each point on the curve corresponds to a vector wind force that can be produced with the appropriate sail orientation. The top views of five little sails decorating the outside of the curve show the sail orientations needed to provide the forces indicated by the smaller arrows. In this and subsequent figures, the apparent wind, $\vec{V}$, is from the south (bottom of the page). In general, the boat's motion means the true wind, $\vec{W}$, is not from the south.

The wind-force curve consists of two clearly different sections. The lower part is a parabola and the upper part is a circle. To explain:

1. The sail's lift-drag characteristics on the lower curve are analogous to those of the centerboard's lift-drag parabola $\vec{F}(water)$. The sail's angle of attack is small because the wind is just skimming past its surface. Just as with the centerboard, a small increase in the sail angle produces a large increase in lift and smaller increase in the drag. The point at the bottom of the parabola corresponds to a luffing sail exactly aligned with the apparent wind. This sail orientation is never efficient.

2. The upper boundary of the shaded area shows the wind force when the angle of attack between the sail and the wind is larger. The sail is no longer acting aerodynamically and the physics is different. It is similar to sailing directly downwind in the sense that the wind on the back side of the sail is not flowing smoothly past the sail.

The point on the top of the Figure 3.15 corresponds to a sail positioned for downwind sailing, extended perpendicular to the apparent wind. The force on the sail is entirely drag with no lift, so $L = 0$.

A simple but imprecise approximation is used to obtain the sail forces on other parts of the upper curve. It is assumed that the force is produced only by a pressure difference, and this pressure difference is the same for all points on the upper curve. In other words, the force is perpendicular to the sail, and the magnitude of the force does not vary with sail orientation. Changing the sail angle changes the direction of the force but not its magnitude. Thus, the upper curve has a circular shape whose radius is proportional to the magnitude of the force.

This approximate form of the wind-force curve introduces errors on the order of 25%, but it is qualitatively faithful to sailboat physics.

3. The "kink" in the curve where the upper and lower pieces meet corresponds to a transition between the two sailing modes. These modes correspond roughly to "impact sailing" on the upper curve and "aerodynamic sailing" on the lower curve. Sailors often want to adjust their sails so that $\vec{F}(wind)$ is at this transition point because sailing at the kink produces maximum lift. This maximum lift-to-drag ratio is called $[L/D](wind)$. The angle between the apparent wind and the force at this special point is denoted v^* in Figure 3.15.

The simplifying approximations of the sail force curve are most outrageous at the kink. For real sails, the kink is smoothed out to a "transition region" between impact sailing and aerodynamic sailing. Sometimes the word "stall" is associated with the transition region.

3.3.6 Step 5: Wind and Water Forces Combined

The next step in the long path to sailboat speeds is a comparison of the forces $\vec{F}(wind)$ and $\vec{F}(water)$. In steady-state sailing, the force needed to push the sailboat is exactly the force supplied by the wind. Since $\vec{F}(wind) = \vec{F}(water)$, superposing these forces on a single graph yields real sailing conditions at points where the curves touch.

Equating the forces essentially solves the problem of sailboat speeds, but there are annoying details. First, one must make sure all forces are measured in the same units. Second, each sailing direction requires a different graphical comparison because the orientation of the curves changes with the sailing direction. After three example directions are considered, the general result is presented.

3.3.6.1 Scaled Units
Only one $\vec{F}(wind)$ curve is shown in Figure 3.15 because only one wind speed was considered. There is a way to avoid drawing a different wind-force curve for each wind speed. For each wind speed, pick the units of force so the upper circle of the wind force curve has unit radius. Then only

one curve is needed. There is a complication to the simplification. For each wind speed, the water force curve must be expressed in the same units. Replacing F_D and F_L with the scaled D and L gives a version of Equation 3.14 that can be compared directly to the wind force.

$$D(water) = \left(\frac{U}{S_0 V}\right)^2 + \left(\frac{S_0 V}{U}\right)^2 \left(\frac{1}{2[L/D](water)}\right)^2 L(water)^2$$

(3.15)

Before proceeding, one should make sure this works for a simplest case of downwind sailing. If the sails are adjusted properly, there should be no lift force downwind. The wind drag was scaled so $D(wind) = 1$. Force equality means $D(water) = 1$, as well, and letting $L(water) = 0$ in Equation 3.16 means

$$D(water) = 1 = \left(\frac{U}{S_0 V}\right)^2$$ (3.16)

This is just a restatement of Equation 2.8, $U = S_0 V$, which describes downwind sailing. Retrieving the previous result is really a check to make sure Equation 3.15 has been properly scaled.

3.3.6.2 Comparing Graphs   A superposition of the wind-force and water-force curves for downwind sailing is shown in Figure 3.15. Intersection points correspond to equal forces. Six different $\vec{F}(water)$ curves are placed on top of the single $\vec{F}(wind)$ curve. Each curve corresponds to a solution to Equation 3.15 with a different downwind boat speed U. Counting up from the bottom, these speeds are $S_0 V \cdot \sqrt{1/5}$ and $S_0 V \cdot \sqrt{2/5}$ and $S_0 V \cdot \sqrt{3/5}$ and $S_0 V \cdot \sqrt{4/5}$ and $S_0 V$ and at the top $S_0 V \cdot \sqrt{6/5}$. The top $\vec{F}(water)$ curve corresponds to a speed faster than is possible because $U > S_0 V$, so it makes sense that this curve never touches the wind-force curve. The next $\vec{F}(water)$ curve describes proper downwind sailing, with $U = S_0 V$. A little sailboat is centered at the top point where the two curves touch. This sailboat shows that proper downwind sailing is achieved with the sail fully extended. There is no lift force so the boat can aim directly downwind.

The slower speed $\vec{F}(water)$ curves also intersect $\vec{F}(wind)$. Little sailboats have been centered at two of these slower-speed intersection

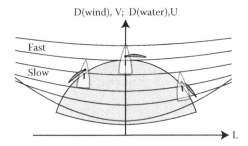

Figure 3.16 Downwind lift and drag of the wind and water displayed in the scaled (L,D) units. The drag axis is also labeled with the U and V because both these velocities are north in this example. The several water-force curves correspond to different boat speeds. Little sailboats have been drawn on three of the intersection points where the wind and water forces are equal.

points. They show that it is possible to sail downwind with a speed less than $U = S_0 V$ by pulling the sail toward the center of the boat so the wind's force is deflected to the side. This action reduces the wind's drag and adds lift. The boat must sail slightly to the side to compensate for the leeway produced by the lift, and this increases the water's drag. Figure 3.16 shows that the boat on the left is sailing at speed $U = \sqrt{4/5} \cdot S_0 V$, and the boat on the right that has pulled sails in even further is a sailing at $U = \sqrt{3/5} \cdot S_0 V$.

Pulling a sail in when sailing downwind is generally a foolish move, but there are occasions when it may be useful. If a boat is about to overrun the stern of a boat ahead, reducing speed is definitely a good idea. Such situations occur more often than one might expect when sailboats are racing around turning marks.

Up to now, the only thing accomplished is an exceedingly cumbersome derivation of the downwind sailing speed. By changing the direction the boat is moving, we can finally calculate boat speed for sailing in other directions.

3.3.6.3 Broad Reach   The water force $\vec{F}(water)$ is changed when the boat sails in a different direction. If the sailboat is sailing northeast instead of north, the $\vec{F}(water)$ curves are rotated because the $D(water)$ axis is parallel to the boat velocity $\vec{U}$, as shown in Figure 3.16. The apparent wind $\vec{V}$ continues to blow from south to north, so the shaded part of Figure 3.15 is unchanged and wind's drag is still north.

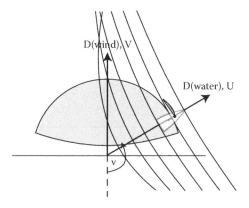

Figure 3.17 Water forces are tied to the direction of the boat's velocity, $\vec{U}$, so this component of the figure is rotated when the angle between the sailing direction and the source of the apparent wind, labeled v in the figure, is not 180°.

Again, a sailor has a choice of sail orientations when sailing on the broad reach. Trimming a sail too tightly will again place the boat on a slower curve. The boat shown in Figure 3.17 has picked the proper sail orientation for maximum speed. The fastest sailing speed is again $U = S_0 V$. The ratio of boat speed to the apparent wind speed has not changed because the upper sail-force curve has an (assumed) circular shape. This simple model says the sail should remain fully extended so the wind force pushes the boat directly forward. No lift is required from the water, and there is no leeway on this broad reach.

Even though the boat speed is the same fraction of the apparent wind speed, $U = S_0 V$, the ratio of the boat speed and the *true* wind speed W is increased. This outcome is not clear from Figure 3.16 because it is related to the apparent wind speed. The correction for the difference between the true and apparent wind is Step 6, Section 3.3.7.

Sailors may view Figure 3.17 as unrealistic, since they know that some lift from the water is required when sailing on a reach. The sailors are right. However, after correcting for the difference between the true wind and the apparent wind direction, the boat shown in Figure 3.17 is actually sailing much closer to downwind than the figure suggests. Sailors know that very little lift is needed when sailing close to downwind, so the error of Figure 3.17 is real, but it is smaller than one would think.

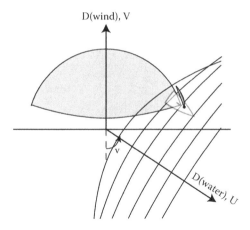

Figure 3.18   Sailing closer to the wind. Fastest sailing requires the sail to be adjusted to produce maximum lift. The ratio of sailboat speed to apparent wind speed is reduced and the intersection point lies on the curve with $U = S_0 V \cdot \sqrt{3/5}$.

3.3.6.4 Sailing Closer to Windward   Further rotation of the water-force curve is required for sailing closer to the wind. For the upwind sailing example shown in Figure 3.18, the boat sails fastest when the wind-force and water-force curves touch at the transition point (kink) on the wind-force curve. For this example, $\vec{F}\,(wind)$ touches $\vec{F}\,(water)$ when the boat speed is a smaller fraction of the apparent wind speed. The maximum speed of the boat, relative to the apparent wind, is now only $U = S_0 V \cdot \sqrt{3/5}$. A small sailboat has been placed at the point where the curves touch. Unlike downwind and broad-reach sailing, the sail is not fully extended, and the wind pushes the boat to the side. Considerable lift is required from the centerboard, and the boat makes some leeway. As in the other examples, the boat will sail slower if the sail is improperly adjusted. Pulling the sail in moves the boat to the circular part of the wind-force curve. Letting the sail out moves the boat to the bottom parabola of the water-force curves. Either adjustment slows the boat significantly. When sailing in this direction, the speed penalty for incorrect sail alignment is more severe.

3.3.6.5 Generalization   For downwind and broad-reach sailing, the boat speed is a constant fraction of the apparent wind speed, given by $U = S_0 V$. For upwind sailing, the ratio of boat speed to apparent wind speed is decreased whenever the apparent wind angle is smaller than the direction to the transition point, or $v < v^*$. (The transition point angle v^* is shown in Figure 3.15.) The example in Figure 3.18 gives $U = S_0 V \cdot \sqrt{3/5}$, but a different angle v would give a different value for U/V.

The general result that includes both upwind and downwind sailing is obtained by replacing the downwind speed ratio S_0 by an angle-dependent speed ratio, $S(v)$.

$$U = S(v)V \tag{3.17}$$

Figure 3.18 shows that fastest upwind sailing is accomplished when a water-force parabola just touches the transition point (kink) on the wind-force curve. This condition is expressed algebraically as a quadratic equation. Combining the quadratic equation result for $v < v^*$ with the condition $S(v) = S_0$ for $v > v^*$ gives

$$
S(v)^2
= S_0^2 \left\{
\begin{array}{ll}
\dfrac{1}{2}\cos(v^* - v)\left[1 + \sqrt{1 - \left(\dfrac{1}{[L/D](water)}\right)^2 \tan^2(v^* - v)}\,\right] & v < v^* \\[4ex]
1 & v > v^*
\end{array}
\right.
\tag{3.18}
$$

Although Equation 3.18 may appear hopelessly complicated, it yields a sensible result. The ratio of the boat speed to the apparent wind speed decreases as the apparent wind angle v decreases.

3.3.6.6 Closest to the Wind In some cases, when a boat sails very close to the wind, the intersection of the wind-force and water-force curves can move to a point on the lower parabola of the wind-force curve. For the iceboat, this part of the curve is important because the maximum value of lift to drag determines the speed. But for sailboats, this portion of the wind force usually corresponds to sailing too close to the wind (pinching). A pinching example is shown in Figure 3.19, where $U = S_0 V \cdot \sqrt{1/5}$ corresponds to relatively slow upwind progress, even though the apparent wind angle v is small. An attempt to sail even closer to the wind will fail because the wind-force and water-force curves will not touch.

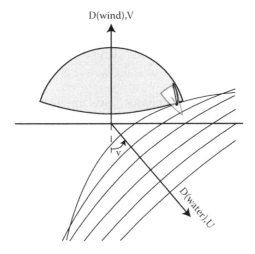

Figure 3.19 Sailing as close to the wind as possible. This boat is "pinching" and would make better progress by sailing at a larger angle, *v*.

3.3.7 Step 6: Sailboat Speed Diagram

The final step is a transformation from $S(v)$, given by Equation 3.19, to the speed diagram, which is a polar plot of $U(w)/W$. Another side trip into the geometry of the velocity triangle of Equation 3.1 is needed to express the boat speed in terms of the true wind. This final step will finish the job.

The velocity triangle is shown again in Figure 3.20. The extra line *d* and two additional distances, *x* and *y* have been added.

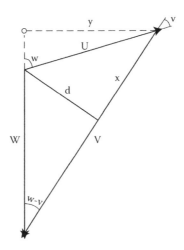

Figure 3.20 Velocity triangle with added distances *x* and *y*.

Using $x = U \cos(v)$, $d = U \sin(v)$, $W^2 = d^2 + (V - x)^2$ and $\sin^2 v + \cos^2 v = 1$ gives the "law of cosines," which generalizes Pythagoras to a triangle without a right angle,

$$W^2 = U^2 + V^2 - 2UV \cos(v) \qquad (3.19)$$

Combining this with $U = S(v)V$ gives a generalization of the relation between the true wind speed, the boat speed, and the speed ratio

$$U = \frac{S(v)}{\sqrt{1 + S(v)^2 - 2S(v)\cos(v)}} W \qquad (3.20)$$

Using $y = U \sin(w) = V \sin(w - v)$, the trigonometric identity $\sin(w - v) = \sin(w)\cos(v) - \cos(w)\sin(v)$ and the definition of the speed ratio $U = S(v)V$ gives another triangle identity.

$$\tan(w) = \frac{\sin(v)}{\cos(v) - S(v)} \qquad (3.21)$$

The desired speed diagram is now obtained from a "bootstrap-like" process:

1. Pick an apparent wind angle v.
2. Find $S(v)$ using Equation 3.18.
3. Use the triangle identity of Equation 3.20 to obtain U/W as a function of v.
4. Use the second triangle identity of Equation 3.21 to replace v with the appropriate w and obtain $U(w)/W$.

In principle, these equations are simple enough that a sailor with a scientific calculator could compute boat speed. A computer program shortens the speed diagram calculation time from hours to less than a second.

3.3.7.1 Basic Example: A Standard Sailboat　The three numbers chosen to characterize the "standard sailboat" are $S_0 = 1$, $[L/D](water) = 10$, and $[L/D](wind) = 3$, which corresponds to the kink angle (see Figure 3.15) $v^* \cong 110°$. These values for the downwind speed ratio, the lift-to-drag ratios, and the kink angle are roughly characteristic of an efficient sailboat of moderate size. The speed diagram obtained from these three numbers is shown in Figure 3.21.

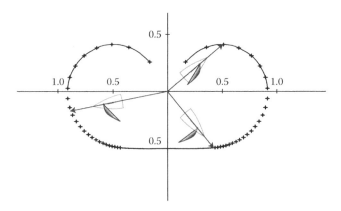

Figure 3.21   A "standard sailboat" speed diagram. The axis scales are values of U/W. The separations between the little crosses correspond to 3° changes in the apparent wind angle. They are not shown at the bottom where they crowd together, being separated by only 1.5°.

As with the iceboat speed diagram, the distance from the origin to any point on the speed diagram's curve gives the ratio of the boat speed, U, to the true wind speed, W. The true wind angle, w, is the angle between a point on the speed diagram and due north (straight up). The direction a sailboat moves can differ a little from the direction that the boat is aimed because of leeway. The diagram is decorated with three little sailboats that show the approximate sail orientation for three example sailing directions.

The standard sailboat speed diagram shows that the downwind speed is half the wind speed, which should not be a surprise. The downwind result (for $S_0 = 1$) was obtained with much less trouble in Chapter 2. Downwind is not the fastest sailing direction. By sailing on a reach with $w \cong 90°$, the standard sailboat achieves its maximum speed that is a little more than 90% of the true wind speed.

The sailboat at the upper right of the diagram illustrates a sailing direction that makes the most rapid progress to windward. For this direction, the speed is about two-thirds the wind speed. However, since the boat is sailing at an angle w that is a little larger than 45°, the windward component of the boat speed is only about 42% of the wind speed. Thus, net progress for upwind sailing is slower than downwind sailing, even though sailing upwind moves the boat faster through the water.

For every sailing direction except downwind, the apparent wind angle v is always less than the true wind angle w. For the standard sailboat shown in Figure 3.21, the comparative values are

- Fastest progress to windward (sailboat at upper right); $w \cong 45°$, $v \cong 33°$.
- A reach, perpendicular to the true wind: $w \cong 90°$. $v \cong 50°$
- At the edge of the flat section (sailboat at lower right); $w \cong 145°$, $v \cong 110°$

These angle differences mean a sailor can think the wind is coming from the bow even when the true wind is more from the stern.

When sailing to windward, a relatively large change in direction is needed to change the apparent wind angle. The opposite is true for downwind sailing, where a small change in sailing direction produces relatively large changes in the apparent wind direction. This is illustrated by the little crosses on the speed diagram, which are separated by 3° increments of the apparent wind direction, starting at $v = 27°$. This standard sailboat cannot sail at an angle smaller than $v \cong 26°$. The crosses are not shown at the bottom of the speed diagram where they crowd together and are separated by a constant 1.5°.

Downwind sailing has a special property when $S_0 = 1$. The flat section of the speed diagram for directions near downwind means one can sail directly downwind at half the wind speed, or one can sail at a 30° angle from downwind and achieve the same downwind progress. When considering a strategy, this gives sailors extra flexibility, as described in Section 10.1.4.

3.3.7.2 Comparison of Speeds   Different sailboats are characterized by different downwind speed ratios S_0, which changes the shape and size of the speed diagram. Speed diagrams for three different model sailboats are shown in Figure 3.22: a slow one with $S_0 = 2/3$, the standard sailboat with $S_0 = 1$, and a fast sailboat with $S_0 = 3/2$. The $[L/D]$ ratios for wind and water are all the same as the standard sailboat values used to generate Figure 3.21.

The fast and slow curves in the Figure 3.22 represent fairly extreme cases. Most sailboats, with the possible exception of some catamarans and light weight skiffs, are characterized by downwind speed ratios between 2/3 and 3/2.

Boat speed is the obvious difference between the three curves in Figure 3.22, which shows that some sailboats can sail on a reach

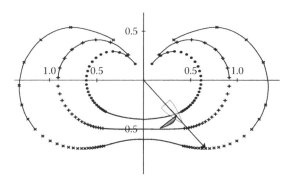

Figure 3.22   Three different speed diagrams for slow $S_0 = 2/3$, standard $S_0 = 1$, and fast $S_0 = 3/2$ sailboats. The lift-to-drag ratios are the same for all three boats.

faster than the wind. Sailboats with $S_0 > 1$ can make better progress downwind by sailing at an angle w that is less than 180°. Sailing directly downwind is a bad strategy for these fast boats because the boat speed so effectively subtracts from the wind speed that "downwind" is nearly "no wind." For the boat with $S_0 = 3/2$, the apparent wind speed for sailing directly downwind is 40% of the true wind speed because the boat speed is 60% of the wind speed. However, when sailing at $w = 135°$ or 45° from downwind, apparent wind speed increases to about 55% of the true wind speed. When sailing in this direction, the apparent wind is from the bow at $v \cong 70°$. The net progress downwind is increased by about 1 part in 6 by sailing at this angle.

Heavier cruising boats and some smaller sailboats are more accurately characterized by the $S_0 = 2/3$ slow boat speed diagram. For these boats, the best downwind course is a straight line.

The difference between sailboat speeds is most noticeable on a reach. When sailing downwind, the fast boat in Figure 3.22 is 1.5 times as fast as the slow boat. On the reach, the fast boat speed is 2.2 times the speed of the slow boat.

When sailing to windward, the speed ratio of the fast to slow boat is again a little more than two to one. However, because the fast boat must deal with a larger shift in the apparent wind direction, it cannot sail at as close an angle to the wind as the slow boat, so the ratio of progress to windward is a little less than two to one.

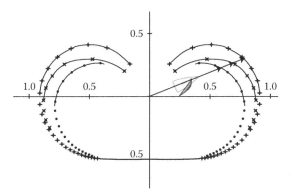

Figure 3.23 A comparison of speed diagrams when lift-to-drag ratios are decreased. The outer curve is the standard sailboat. The center curve results from a decreased lit-to-drag ratio from the water, and the inner curve results from a reduced lift-to-drag ratio from the wind.

3.3.7.3 Comparisons of Lift-to-Drag Ratios The lift-to-drag ratios of the wind and water are important for upwind sailing. In Figure 3.23, the standard sailboat speed diagram of Figure 3.21 (outer curve) is compared with diagrams where the water's lift-to-drag ratio is reduced by a factor of three (middle curve) or the sail's lift-to-drag ratio to two-thirds of its initial value (inner curve). These alternative speed diagrams could result from neglecting to lower a centerboard or having a sail that is too "baggy." The lift-to-drag ratios have essentially no effect on downwind sailing.

The results shown in Figure 3.23 are sensitive to the initial choice of lift-to-drag ratios. However, the basic conclusion is always the same. Reducing the lift to drag of either the wind or the water always slows upwind progress.

3.4 Why Is Sailing Upwind So Complicated?

Is it really necessary to draw so many diagrams and write so many formulas to describe sailing? Sadly, I have been unable to find a shortcut path to the speed diagram. In fact, sailing is really more complicated than all the diagrams and formulas suggest. A good deal of cheating was needed to obtain the shapes of the speed diagrams that are the ultimate accomplishment of this section. It is worth remembering the sacrifices of accuracy that were needed to obtain the results. A few of the shortcuts are listed here.

- The quadratic approximation means that the sailboat speed is proportional to the wind speed. However, neither the wind nor the water forces ($\vec{F}(wind)$, $\vec{F}(water)$) are exactly proportional to the square of the wind speed. If the ratio $U(w)/W$ is accurately measured or calculated, one finds that it often decreases with wind speed, so a separate speed diagram is really needed for every wind speed. Real examples of the relative decrease in boat speed are shown in Figure 2.9. For some sailboats, this means $S_0 > 1$ in light winds but $S_0 < 1$ on windy days. The tactical significant of this change in S_0 is described in Chapter 10.

- Many sailboats deploy spinnakers for downwind sailing that increase $S(v)$ for large angles v where the spinnaker can be used. A boat equipped with a spinnaker really has two speed diagrams. Many sailboats use different sized jibs in different wind conditions. A change in sails means a change in the speed diagram. When the wind is so strong that not all of the sail's power can be used because of the danger of capsize, the speed diagram is altered.

- The wind force $\vec{F}(wind)$ sketched in Figure 3.15 is an idealization. The upper part of the curves is not really part of a circle and the transition between impact sailing and aerodynamic sailing is not really a sharply defined kink. In principle, this error could be corrected by using measured curves for $\vec{F}(wind)$. Although the details are different, the procedure would be the same and the sailboat speed can still be determined graphically by finding points where the wind and water forces are equal.

- The force on a sail is not entirely due to pressure. As the wind slides by a sail, there is a side force due to the air's viscosity. This side force must be countered by lift from the centerboard. Also, since the upper curve of $\vec{F}(wind)$ is not exactly a circle, a sail should often be trimmed more tightly on a broad reach. This gives an addition side force that must be countered by a centerboard.

- When there is enough wind to heel a boat, all the forces are changed. Matching the altered wind and water force of a heeled sailboat changes the speed diagram.

These and many other complications mean the speed diagrams shown in Figures 3.21–3.23 are not the speed diagrams of any real sailboats. For example, the speed diagrams derived here all show that the most efficient upwind sailing corresponds to a true wind angle that is a little more than 45°. In the real world, some sailboats can sail efficiently at slightly smaller true wind angles. Which oversight is responsible for this difference between theory and practice is not clear to me.

4

TIPPING, TORQUES, AND TROUBLE

4.1 Roll, Pitch, and Yaw

Boats rock, bob, and weave in complicated ways as they respond to the wind, the waves, and the person holding the tiller. To make sense of this, three kinds of angular motion are distinguished. Roll is side-to-side tipping. Extreme roll leads to capsize. Pitch corresponds to rotation where the bow lifts and falls. Extreme pitch leads to an especially exciting summersault version of capsize called a "pitch poll." Yaw is a change in direction. Often one thinks of yaw as the unintentional variation in direction caused by waves and the wind. Too much roll, pitch, and yaw results in discomfort and sea sickness.

A boat that rotates too easily is unstable and uncomfortable. Designing a sailboat that stays upright but still moves easily through the water is a practical problem with an ancient history. Long ago, Archimedes investigated the stability of some simple floating shapes. Roughly 2,000 years later Leonhard Euler and a less famous pioneer of nautical engineering, Pierre Bouguer, explained the basic physics of boat stability, which is the foundation of our present knowledge.

4.2 Torques

For angle changes, torques play the role of forces. A torque twists or rotates something without changing its position. The sailboat is subject to torques when the wind pushes the sail to one side and the water pushes back with an equal and opposite force on the centerboard (or keel) and hull. This torque will capsize the boat unless there is an opposing torque of equal magnitude.

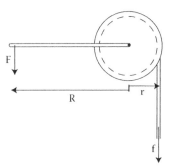

Figure 4.1 A winch showing opposing torques $\tau_1 = F \times R$ and $\tau_2 = -f \times r$.

4.2.1 Winch: A Simple Example

A winch is an example that may be too familiar to exhausted crew members on large sailboats. The winch sketched in Figure 4.1 is a cylinder that can rotate on its fixed axis.

A crew member supplies a torque by pushing with a force F on a bar attached to the cylinder. The wind supplies an opposing torque by pulling with a force f on a line wound around the cylinder. The general formula for the magnitude of a torque

$$\text{Torque} = \text{Force} \times \text{Distance}$$
$$\tau = F \times R$$
(4.1)

allows one to evaluate the torques applied to the winch. If the bar has length R, the crew's torque is $\tau_1 = F \times R$. The wind's torque is $\tau_2 = -f \times r$, where r is the cylinder radius. The negative sign for the wind's torque means it rotates the cylinder in the opposite direction. The cylinder will not rotate if the two torques cancel, so requiring that $\tau_1 + \tau_2 = 0$ gives

$$f = \frac{R}{r} F$$
(4.2)

The crew member's force, F, is magnified by the winch. For the example in Figure 4.1, Equation 4.2 shows that the magnification factor is $(R/r) \cong 4$, which can be a great help when trimming a large sail on a windy day.

4.2.2 More General Torques

There are (at least) four ways that sailboat torques generalize the torques applied to the winch.

1. The sailboat can rotate in three different directions. A separate torque determines roll, pitch, and yaw.
2. For the winch, the distance in Equation 4.1 is between the point the force is applied and the rotation axis. A sailboat has no obvious rotation axis, but it is generally acceptable to use the sailboat's "center of mass" as the reference point.
3. For the winch, the forces are perpendicular to the distances. When forces and distances are not perpendicular, the formula is modified. A horizontal force is multiplied by only the vertical part of the distance, and a vertical force is multiplied by only the horizontal part of the distance.
4. The forces on a sailboat are not applied at one point. For example, all parts of a sail contribute to the total sail force. The torque formula can still be applied if the force is considered to act at its "center of effort."

For a sailboat, one must consider torques produced by the wind, water, and buoyancy. The total torque for each rotation direction must vanish if the boat is to sail steadily. For high-performance sailboats, achieving balanced torques can be delicate and difficult.

4.3 Centers of Mass, Buoyancy, and Effort

Gravity pulls on every part of the boat. Wind pushes on the entire sail. Water's buoyant force pushes up on all submerged surfaces. These distributed forces are equivalent to forces acting at a single point. Knowing the positions of these special points simplifies the calculations of the torques.

4.3.1 Center of Mass

The most familiar "center" is the center of mass. Simple experiments (in principle) can determine the center of mass of any object, including a sailboat. Suspend a sailboat from a single point. A line drawn directly down from the suspension point passes through the center of mass. It doesn't matter if you lift the boat by the bow, the stern, or the mast. All lines descending from the suspension points pass through the same center of mass. Sailboat symmetry means the center of mass lies in the plane that separates port from starboard.

Although gravity pulls down on every piece of a sailboat, all the forces act as if they were pulling on the center of mass with a force Mg, where M is the total mass and $g = 9.8$ m/s^2 is the acceleration of gravity.

4.3.2 Center of Buoyancy

Just as gravity pulls down, buoyancy keeps a boat from sinking by pushing up with a force that is equal but opposite to gravity. Buoyancy pushes up on all submerged parts of a boat, but these forces act as if they were pushing at a single point. That point is the center of buoyancy.

When a boat is placed in water, it sinks down and moves water away. The mass of this displaced water equals the boat's mass. That is Archimedes principle. The center of buoyancy is the center of mass of this displaced water. Although the boat's center of mass is fixed, the center of buoyancy moves as a boat tips. A sailboat is stable when tipping causes the center of buoyancy to move in a direction that stops the tipping.

4.3.3 Center of Effort

Wind pushes on a sail. The total wind force can be represented as if it acted on a single point. That point is the center of effort. A simple example is a mainsail shaped like a right triangle. If the pressure is uniform over a sail's surface, the center of effort is one-third of the way up from the boom to the top of the sail (the "head"), and one-third of the way back from the mast to the end of the boom. If the pressure is nonuniform, the total force can still be represented by a center of effort, but its position will be different.

There is also a center of effort for the water's horizontal force on a centerboard (or keel). This center of effort is centrally located on the centerboard.

4.4 Catamaran

The catamaran is the first application of torques because its geometry makes estimates relatively simple. A typical catamaran sailboat has two narrow canoe-like hulls that are separated by about half a boat length. The mast is centered in a structure rigidly connecting the two hulls. Over the ages, variations of the catamaran have been invented many times and in many places. The word "catamaran" [kattumaran = tied logs] comes from the Indian Tamil language. Polynesians colonized much of the Pacific in catamaran-like doubled canoes.

4.4.1 Catamaran Roll and Capsize

Normally, catamarans don't tip much, unless they tip over. The forces jeopardizing a catamaran on the edge of stability for roll are shown in Figure 4.2. For the boat in (a), the left hull is about to leave the water. Any increase in sail force produces the tipping shown in (b).

The rotational stability is determined by three torques (assuming the mass of the very skinny sailor is ignored). The first torque τ_1 is produced by the wind's horizontal force F on the sail. The center of effort is a vertical distance h_1 above the center of mass, so

$$\tau_1 = -F \times h_1 \qquad (4.3)$$

The negative sign applies because this torque rotates clockwise.

The second torque τ_2 is produced by the equal and opposite horizontal force of the water on the catamaran hull that keeps it from sliding sideways. (For simplicity, the hull was taken to be wedge shaped, so no centerboard is needed.) The center of effort of the water force is a vertical distance h_2 below the center of mass. Thus,

$$\tau_2 = -F \times h_2 \qquad (4.4)$$

The negative sign indicates another clockwise torque.

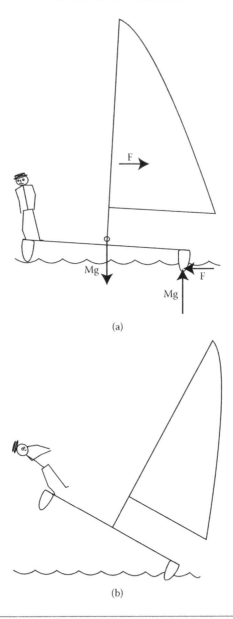

(a)

(b)

Figure 4.2   (a) A sailboat stays balanced when the sum of the applied torques vanish. (b) When the torque from the wind and water exceeds the compensating torque supplied by buoyancy, capsize results.

The third torque is produced by the buoyant force, which is equal to the force of gravity, Mg. If the catamaran is barely balanced, this force is applied only to the right hull section. Half the boat width (or beam), $B/2$ is the distance to the center of gravity. Thus,

$$\tau_3 = Mg \times \frac{B}{2} \qquad (4.5)$$

The positive sign means the buoyancy torque is counterclockwise.

The boat will be stable only if the sum of the three torques vanishes.

$$\tau_1 + \tau_2 + \tau_3 = 0 \qquad (4.6)$$

Setting the sum of torques equal to zero yields a maximum value for the force of the wind.

$$F(\text{max}) = \frac{MgB/2}{h_2 + h_2} \qquad (4.7)$$

A wind force any larger than $F(\text{max})$ will tip the boat over. Equation 4.7 makes intuitive sense. In order to withstand a larger wind force, the catamaran should be wider (larger B) or heavier (larger M). If the mast is taller (larger h_1) or the water force is applied to a deep board (larger h_2), the $F(\text{max})$ will be smaller.

Rough estimates of the forces and wind speeds are given here for a catamaran with properties similar to a Hobie-17 (see Figure 4.9), $L \cong 5.2$ m, $B \cong 2.3$ m, and $M \cong 143$ kg. The mast is about 8.2 m high, so $h_1 + h_2 \cong 3$ m is a reasonable rough approximation, which yields $F(\text{max}) \cong 540$ N. Any greater wind force would capsize the boat.

The maximum sail force $F(\text{max})$ can be used to estimate the wind speed needed for capsize. Assuming the wind is coming from the side, the wind force is the drag force. Using Equation 2.6 again,

$$F = \frac{C_D}{2} \rho \cdot A \cdot V^2 \qquad (2.6)$$

The drag coefficient for downwind sailing was roughly estimated to be $C_D \cong 4/3$. For wind coming from the side, the sails would be adjusted to make a considerably smaller (C_D). A reasonable (but very approximate) estimate is $C_D \cong 2/3$. The sail area on a Hobie-17 is about 16 m². Using these estimates, the boat would capsize when $V > 9$ m/s.

This result seems unrealistic because 10 m/s is only a Fresh Breeze. The error lies in the neglect of the crew member's weight. If the stick figure man in Figure 4.2 has $m = 75$ kg, and if he sits in the appropriate extreme left position, his mass would contribute an additional balancing torque. Because the crew member with half the boat mass is the full boat width B from the submerged hull section, his restoring torque is equal to that of the boat itself. The doubled torque means the maximum wind could be increased by $\sqrt{2}$, and the boat could remain balanced for $V < 13$ m/s. An apparent wind speed of 13 m/s is considerably more intimidating.

The crucial role of the crew for catamaran balance is typical for many small sailboats. The proper position of the crew in a Fresh breeze or greater wind is essential for boat balance. If the crew falls off the boat and the sails are not released, the boat will probably tip over.

Catamarans do not commonly tip over sideways. When the boat starts to tip, the sail can be released until it luffs, which reduces the force and avoids an embarrassing event. There is still an advantage for a heavier crew in strong winds. If a bigger crew member makes luffing unnecessary, the wind won't be wasted.

4.4.2 Catamaran Pitch

For a catamaran, pitch rather than roll can be the crucial rotation. The story for pitch starts out in a similar manner. Assume the apparent wind is still coming from the side. Rotating Figure 4.2 by 90° means the sail's lift force F_L replaces the drag, as shown in Figure 4.3. As with roll, the center of buoyancy moves forward as the boat starts to tip. If the sail force is too large, the boat will topple forward. As with roll, pitching becomes unstable when apparent winds are strong. The lift force is typically larger than the drag force when sailing in this direction and a catamaran is not that much longer than it is wide, so instability to pitch is a real possibility for some catamarans.

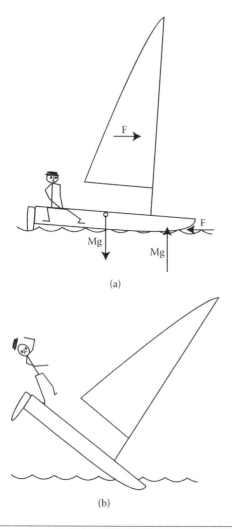

(a)

(b)

Figure 4.3 For pitching motion in (a), uncompensated torques along a different axis produce capsize with a different geometry, shown in (b).

The more dramatic pitch poll capsize has some surprises. As the wind's torque increases and the center of buoyancy moves forward, the bow can plunge, as shown in Figure 4.3b. As soon as this happens there is an abrupt increase in the drag force from the water. It suddenly becomes larger than the sail's force F.

$$F(hull) \rightarrow (BIG) \qquad (4.8)$$

When this happens, forces are unbalanced and the resulting (negative) acceleration a of the center of mass is obtained from

$$(M+m)a = |F_L| - |(BIG)| \tag{4.9}$$

At the same time, the torque from the water is increased

$$\tau_2 \rightarrow (BIG) \times h_2 \tag{4.10}$$

Here h_2 is again the vertical distance between the point where the (BIG) force is applied and the center of mass. Because (BIG) is very large, the boat tips and moves the center of mass up. The increased vertical distance h_2 yields a larger and often irresistible torque.

A sudden increase in torque combined with a sudden backward acceleration of the boat produces a catapult-like action, giving the crew a spectacular but brief experience that some find exhilarating.

Single-hulled boats are too long and narrow to allow the sort of pitch poll that can be experienced in a catamaran. However, a diving bow is seldom a happy event. For many sailboats, it can lead to a complicated capsize involving both rolling and pitching. It is quite a challenge to control a sailboat with a submerged bow, especially if the dive is deep enough to lift the rudder out of the water. Steering becomes impossible.

4.5 Iceboat

Iceboaters who wish to stay off the ice must perform similar balancing acts by reducing sail pressure when the torques become too large. There are some differences. Since two runners stay on the ice when the third runner lifts (as is happening in Figure 4.4), only one rotation is relevant. The rotation axis is determined by the line connecting the front runner to the leeward runner that remains on the ice. To see if the iceboat will tip, it is simpler to calculate torques with respect to this line. It is analogous to the rotation axis on the winch. With this choice, there are only two torques. The first is the wind force applied at the sail's center of effort. The opposing second torque is produced by the weight of the boat plus crew acting at the center of mass. As with the sailboat, when the wind is so strong that its torque cannot be canceled by gravity, the result can be unpleasant.

Figure 4.4   Iceboat on the verge of instability. (Photograph by Stéphane Caron. With permission.)

4.6 Monohull

Most boats are narrow compared to a catamaran, so capsize is potentially a more serious threat. Some boats counter this threat with a weighted keel. There is a big difference between sailboats with keels and sailboats that lack keels. It is virtually impossible for a heavy keel sailboat to capsize. No matter how far the boat tips, there is an opposite torque that restores the boat to its upright position. This is illustrated, and slightly exaggerated, in Figure 4.5.

The heavy keel on the boat on the left of Figure 4.5 moves the center of mass close to the end of the keel. Even when the boat is tipping at 70°, there is a large restoring torque, given by $\tau = Mg \times R$, where R is the horizontal distance between the center of mass and the center of buoyancy. The similar boat on the right that lacks a keel is on the verge of instability

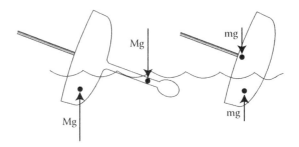

Figure 4.5 A comparison of gravity and buoyant forces for boats with and without a weighted keel. For both boats, gravity acting at the center of mass is indicated by the arrow pointing down. The buoyant force acting on the center of buoyancy is indicated by the arrow pointing up.

when it is tipped 70° because the center of mass lies directly above the center of buoyancy, corresponding to $R = 0$ and no restoring torque.

When a boat is tipped at 70°, the wind force on the sail is much reduced, so one would expect the restoring force on the keel boat to save the day. This is not the case for centerboard boats.

For small centerboard sailboats, the restoring torque is initially proportional to the tipping angle, but when the boat tips around 30° or more, the center of mass starts to move over the center of buoyancy, reducing the restoring torque. Any small boat sailor who has capsized is familiar with this unpleasant effect. Once a sailboat tips too far, little can be done to save the ship. A boat in this unfortunate position is shown in Figure 4.6.

Figure 4.6 Trouble. (Photograph by Sally Snowden. With permission.)

The maximum distance between the center of mass and the center of buoyancy is roughly one-quarter the boat's maximum width (beam). So narrow boats produce smaller restoring torques. This agrees with the commonsense observation that it is easier to capsize a canoe than a raft. The restoring torque is also larger for a flat-bottomed boat than one that is curved, and a boat with a lower center of gravity is more stable. That means you should not stand up in a canoe. The ultimate in watercraft instability can be seen in logrolling contests. Placing a sail on a log does not produce comfortable sailing.

For a small sailboat, it is easier to measure the restoring torque than it is to calculate. A docked boat can be capsized with a line tied to the top of the mast. Pulling the line in a direction perpendicular to the mast exerts a torque that is approximately the product of the force and the height of the mast. If the angle between the line and the mast is θ instead of 90°, the applied torque is scaled by $\sin(\theta)$. As the boat types, the torque first increases and then decreases as the mast approaches horizontal.

4.7 Staying Upright

When the wind is strong enough to make heeling a serious problem, sailors have two choices. They can either decrease the torque produced by the sail or increase the restoring torque that keeps the boat upright. Even if capsize is avoided, sailboats generally sail poorly when the heel angle is large. The sails of a radically heeled boat present a smaller cross section to the wind and some of the sail's lift is directed down into the water. This is not a good direction. When the centerboard or keel is also tipped, it is less effective in preventing leeway. For example, the wake behind the overheeled *Annie* in Figure 4.10 suggests significant leeway.

4.7.1 Limiting the Sail's Torque

Because the sail torque is force times average height, it makes sense to decrease both the sail force F and the height. This can be done by "reefing" the sail so the bottom portion is tied to the boom. If the bottom 10% of a triangular sail is eliminated, the area will be multiplied by $(0.9)^2$, and the center of effort will be moved 10% closer to the boom, leading to a torque reduction of around 25%. If reefing is not possible, the sail angle can be adjusted so that it luffs, and the wind pushes effectively on only a portion of the sail. If possible, it is best to make the top part of the sail luff, since that is the sail section that supplies the most torque.

Reefing is not common on small sailboats. However, many sailboat classes allow different size jibs that make it possible to tailor sail area for the wind.

One could design a sailboat with a short mast and a long boom. The decreased sail height would decrease the torque. However, there is a price to pay. Short sails are less efficient because the wind can more easily slide over the top of a short sail.

A flatter sail can improve sailboat stability and speed in heavy winds. As a simple example, consider a case where the apparent wind is abeam. For this case, the sail's drag and lift forces play clearly defined roles. The drag F_D heels the boat and the lift F_L powers the boat ahead. Assuming the sail is trimmed to maximize the ratio of lift to drag (denoted $[L/D](wind)$), the lift force is

$$F_L = F_D \cdot [L/D](wind) \qquad (4.11)$$

A flatter sail produces a larger $[L/D]$ $(wind)$ and a fuller sail increases F_D. The "best" sail is the choice that maximizes the product so the lift is as large as possible.

In heavy winds, too much drag force will capsize the boat. If the drag force cannot be made large, the best choice is the flatter sail that maximizes the ratio of lift to drag while keeping the drag force in check. Similar considerations apply for sailing at other wind angles. In some cases, sail trim can adjust sail shape and make the sail flatter as the wind increases. However, the adjustments are limited, especially for high-tech sails with very little stretch. Because different sail shapes are best in different winds, sailors have a plausible justification for spending even more money on their hobby.

4.7.2 Increasing the Restoring Torque

Sailors who wish to go fast prefer not to decrease their sail force. Instead, attempts to increase restoring torques are the first line of defense in a

Figure 4.7 Strapping the feet down allows a person to move his center of mass beyond the edge of the boat. (Photograph from MC class Web page.)

strong wind. The methods sailors have devised to balance a boat are varied and imaginative. Sometimes they are precarious and uncomfortable.

Sitting on the windward edge of the boat, as the catamaran stick figure did in Figure 4.2, is good. Placing crew members' legs under straps so they can extend beyond the edge without falling off (hiking), as shown in Figure 4.7, is better. Attaching crew members to lines connected to the mast, as shown in Figures 4.8 and 4.9, is best. These and other mechanical aids allow crew members to move their centers of mass beyond the edge of the boat, and thus exert greater torque. Sailors in these unusual positions who do not want to go swimming

Figure 4.8   The crew on this Flying Dutchman is supported by a cable attached high on the mast. As the wind increases the crew will move out farther. (Photograph by Sally Snowden. With permission.)

must respond quickly whenever the wind changes or the helmsman makes a mistake.

The Flying Dutchman sailboat shown in Figure 4.8 provides a good example of the importance of crew placement. The boat mass is 165 kg and the beam (maximum width) is 1.7 m. Moderate heel can move the center of buoyancy roughly half a meter from the center of mass. The torque produced by tipping alone is roughly 800 N/m. The center of mass of a Flying Dutchman crew member on a trapeze can be 1.8 m from the boat's center of mass. If the crew member's mass is

Figure 4.9 The Hobie-17 catamaran can support a large sail and move rapidly because the sailor can supply such a large torque. (Photograph by Richard Olson. With permission.)

80 kg, the torque from the crew member is 1400 N/m. The single crew member is much more important for keeping the boat upright than the boat itself. If some coordination failure causes this crew member to fall into the water, one can expect capsize.

An even larger torque is produced by the sailor in the Hobie-17 shown in Figure 4.9. The "wings" on the edge of the boat and the line from the mast allow this sailor to fly about 3 m from the center of the boat.

One can also maintain boat balance by moving things other than the crew. "Sandbagger" yachts were popular around the 1850s. Extra ballast consisting of many bags of sand or stone, each with a mass of 20 kg or more, was hauled on board. Every time the yacht tacked, the sand bags were dragged to the other side of the boat. These were the extreme sailboats of the day, and they produced a truly strenuous form of sailing. Some of these yachts have been preserved. An exceptional example is the centerboard yacht *Annie*, launched in 1880 and shown in Figure 4.10. This was the first boat acquired by the Watercraft Collection at Mystic Seaport in Mystic, Connecticut, in 1931. It is rigged and afloat for Mystic Seaport Museum visitors to view. The *Annie* is about 8.8 m long, but its sails extend far beyond the ends of the boat. The distance from the end of the bowsprit to the end of the boom is about 24 m. The sail area is 122 m², which is more than triple the sail area of a typical modern sailboat of comparable length. As it is sailed today, the *Annie* uses water containers instead of sandbags as the moveable ballast.

A high-tech version of a sandbagger allows machines to tilt the keel from side to side so the underwater mass can supply the needed torque even though the boat remains nearly level. This technology is not cheap.

4.8 Steering and Helm

The third rotation axis of the roll-pitch-yaw trio determines changes in the boat's sailing direction. The rudder steers the boat because it exerts a torque, pushing the stern to starboard or port. The rudder's torque is only part of the story because the sail and the hull (including keel or centerboard) can also exert steering torques. A simple example, illustrated in Figure 4.11, compares two stern views of a boat sailing downwind. For both the upright and heeled boat, the center of effort of the sail pushing forward is denoted by an X and the center of effort

Figure 4.10 Sandbagger *Annie* (© Mystic Seaport, Photography Collection, Mystic, CT, #1949.843. With permission.)

for the hull pulling back is denoted by another X. If the boat is not heeled, water's center of effort is directly below the center of mass. It exerts no steering torque. However, the wind's center of effort is about one-third of the way out on the boom, so the torque (sail force times one-third of the boom length) is relatively large. To sail straight, the rudder must be angled so its torque cancels the sail's torque. Deflecting the rudder increases its drag. In this example, lift from the rudder is needed to prevent a turn to port and sailing into the wind. The boat has "weather helm." Its opposite is "lee helm."

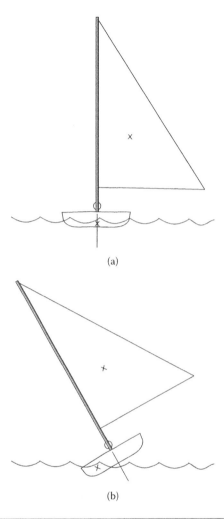

(a)

(b)

Figure 4.11   Stern view of downwind sailing. The X on the sail marks its center of effort pushing forward. The X on the hull is the center of effort (or resistance) pulling back. The unheeled boat in (a) has a large weather helm. Heeling the boat in (b) decreases the horizontal distance between the X's and the steering torque.

For this downwind example, the weather helm and associated rudder drag can be minimized by heeling the boat to windward. Heeling moves the sail's center of effort closer to a position directly over the center of mass, denoted by the small circle in Figure 4.11. Heel also shifts the position of the hull's center of effort so its torque no longer vanishes. Although an unreasonably large heel would be needed to completely eliminate the weather helm, the reduction shown in

Figure 4.11 can produce more rapid sailing. Possible additional advantages of this heel are the higher placement of the sail into a stronger wind and a possible decrease in hull drag because of a reduced "wetted surface." These additional advantages are only speculations. There is an alternative theoretical advantage in having a sail close to the water because it is harder for wind to flow under the sail. Also, hull drag depends on much more than just the wetted surface.

Except for downwind sailing, some weather helm is desirable because the lift force on the rudder enhances the overall lift-to-drag ratio of the boat, as illustrated in Figure 4.12. The arrows show the separate forces on the sail, hull (including centerboard), and rudder. The sum of the forces and the sum of the torques are both zero, so the boat is "balanced" and sailing on a steady course. The boat's leeway has been exaggerated a little in the arrow labeled with U that shows the boat velocity. The leeway means that the rudder's force has a lift component even though it is aligned with the boat's axis. If the rudder is released, the boat will turn into the wind because the dotted line projected from the sail's center of effort is aft of the hull's center of effort.

If the boat in Figure 4.12 were to heel significantly, the sail's center of effort would move out (toward the bottom of the page) and its dotted projection would move farther aft. This would increase the torque and the weather helm. Rudder deflection would be needed to counter the effect of heeling, and this would increase the drag. Most sailors know that excess weather helm is often caused by excess heeling. However, when it is blowing hard, it is not always easy to sail a boat "flat" enough to limit the helm.

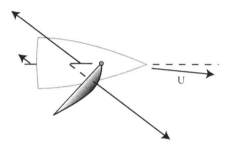

Figure 4.12   Forces on the sail, hull, and rudder for steady sailing. The sailboat velocity with exaggerated leeway is the arrow labeled U.

The weather helm often becomes too large on a reach. Weather helm can be reduced by moving the sail force forward or moving the hull force backward. If the centerboard is pivoted, raising it part way also moves it aft and decreases the weather helm.

4.9 Dynamics

4.9.1 Moment of Inertia

Torques cause rotations in the same sense that forces cause motion. Stated in a more fussy way, torques produce angular accelerations just as forces cause ordinary accelerations. The rotational analogue of Newton's law $\vec{F} = m\vec{a}$ is

$$\tau = I\alpha \tag{4.12}$$

Here, τ is the total torque, I is the moment of inertia, and α is the angular acceleration. The moment of inertia I is like a mass. Heavy objects are difficult to accelerate, and angular acceleration is reduced for objects with large moments of inertia. The moment of inertia of a sailboat is roughly the product of its mass multiplied by the square of its "average" size. This average depends on the boat's shape and mass distribution.

The equation $\tau = I\alpha$ doesn't tell the whole story because there are really three different moments of inertia, one for each type of rotation (roll, pitch, and yaw). To find the different moments of inertia, draw the rotation axis through the boat's center of mass. The contribution to I from each bit of mass is given by the mass multiplied by the square of the distance from the rotation axis. This relationship shows that a long thin boat would have a small moment of inertia for roll, and a much larger moment of inertia for pitch and yaw. Typically,

$$I(roll) \approx \frac{1}{2} M \left(\frac{B}{2} \right)^2 \tag{4.13}$$

$$I(pitch) \cong I(yaw) \approx \frac{1}{3} M \left(\frac{L}{2} \right)^2 \tag{4.14}$$

Here, B is the beam (or width), L is the length, and M is the mass of the boat.

Different parts of the boat add individually to the moment of inertia. For example, the contribution of the mast to $I(roll)$ or $I(pitch)$ is roughly one-third its mass multiplied by the square of its height. The mast contributes much less to $I(yaw)$ because all of the mast is quite close to the vertical rotation axis associated with yaw. The position of the crew also makes a small difference. When a crew member of mass m moves from the center to the end of the boat, the moment of inertial for roll is not changed noticeably, but the changes in the other two moments of inertia are approximately

$$\Delta I(pitch) \cong \Delta I(yaw) \approx m\left(\frac{L}{2}\right)^2 \qquad (4.15)$$

This change can be significant on lightweight sailboats.

In some classes of sailboats, one often sees crew members sitting close together near the center of the boat. The clustering may be an attempt to minimize the moment of inertia for pitch. This reduced moment of inertia can help a boat sail up and over waves rather than cutting through them. This may or may not be a good thing. When a boat is subjected to fairly large waves, the rocking of the boat can whip the mast back and forth at a significant speed. The resulting sail motion (especially near the top) modifies the wind speed and direction relative to the moving sail. Sail pressure develops best when the wind is steady, so a rocking motion can slow a sailboat. For some sailboats, the advantage of sitting snugly side-by-side when sailing upwind may make more difference for wind resistance than it does for changes in the moments of inertia.

4.9.2 Resonance

Whenever a boat tips a little, there is a restoring buoyant force that keeps the boat upright. When someone hops on the bow of his boat, the front sinks until the extra buoyant force from the bow cancels out the additional weight. However, the moment of inertial for pitching means that once the motion starts it takes some extra force to stop it. The boat will first tip too far, and it can overshoot again when returning to equilibrium. Sometimes the bouncing can repeat and last for a few seconds. This motion is roughly analogous to that of a pendulum, but a pendulum is different because it oscillates back and forth many times before it stops. The boat's oscillation damps out quite quickly because the moving water rapidly absorbs the boat's energy. Despite the strong damping, there is a roughly defined natural resonance frequency for the pitching motion of a boat. In principle, there is also a resonance for rolling. The roll resonance is often damped so quickly by the centerboard or keel and rudder that it cannot be noticed. The resonant frequencies are easier to measure than to calculate. Because water must move out of the way as a boat rolls or pitches, the water does more than just stop the motion. It also increases the effective moment of inertia.

On rare occasions, the frequency of water waves passing a boat can coincide with a boat's resonant frequency. The waves can feed energy into the motion. It is as if someone repeatedly hopped onto the boat at just the right times to amplify the pitching or rolling motion. This amplification can be annoying, and it slows the boat's progress. In principle, the amplified motion could be suppressed by a change in resonant frequency accomplished by moving the crew.

4.9.3 Instability

As the wind increases, possible instabilities for both upwind and downwind sailing multiply. On heavy wind days, sailors (like me) continue to discover new ways do things wrong. Only one example is described here. Assume a boat is sailing downwind and tilted to windward (as in Figure 4.11b) to minimize weather helm.

A wind, wave, or crew action moves the mast toward a vertical position, as in Figure 4.11a. Unless the skipper responds quickly, the boat will turn to port in response to the new increased weather helm. If the boat turns quickly to port, the centerboard will exert a lift force (to the left in Figure 4.11). This lift is needed make the boat move along its altered direction to port. The torque associated with this lift increases the heel to starboard, amplifying the original change and endangering the boat's balance.

This is an example of "positive feedback" where a change in boat orientation is amplified by physical forces. Regardless of its origin, positive feedback requires quick action or the instability can be disabling.

Sometimes the forces of instability have a simple explanation, but more often they are quite complicated and involve a number of different effects. Many mechanisms have been proposed to explain the positive feedback associated with different types of sailing catastrophes.

Sometimes waves play an important role. When three complex systems (the wind, the boat, the water) are coupled, any simple explanation of the positive feedback may well be an oversimplification.

4.10 Upright Mast

Torques appear in a variety of other forms on a sailboat. The torques associated with keeping the mast in place is just one example, but it is obviously an important example. The sail forces can be large, so cables (typically stainless steel) are used to keep the mast rigid with respect to the boat. In the simplest arrangement, three cables are connected high on the mast and anchored to the hull. A forestay connects to the bow and two shrouds are anchored to the sides some distance behind the mast. Failure of any cable produces dismasting.

Consider an MC sailboat like that shown in Figures 4.7 and 4.13 that is sailing directly downwind in a Fresh Breeze. The sail area is 12.5 m. The boom is a little more than half a meter above the deck. A reasonable estimate of the center of effort (roughly one-third of the way up the sail) is 3 m above the deck. Applying Equation 2.6 for the downwind force gives $F \cong 250$ N. The corresponding torque (force times distance) is $\tau \cong 750$ N – m. The shrouds are attached to the deck about 1/3 m behind the mast. If each shroud supplies half the torque, the force f on each shroud is obtained by equating torques

$$2f \cdot \frac{1}{3}\text{m} \cong 750 \text{ N/m} \qquad (4.16)$$

This is a large force, $f = 1250$ N. It is larger than the force on the sail for the same reason that forces are magnified by a winch. The shrouds are fairly thin cables, so the strength of materials is a concern when f is more than 1,000 N. The cables have a diameter of about 1/3 cm.

Figure 4.13   Dismasted, but not dismayed? (Photograph from MC class Web page.)

Because they are made from twisted wires, the effective cross section is about $8 \times 10^{-6}\,\mathrm{m}^2$. Thus, the force per unit area on each is

$$\frac{f}{A} = 1.5 \times 10^8 \, \frac{\mathrm{N}}{\mathrm{m}^2} \tag{4.17}$$

Even stainless steel will stretch if the force is strong enough. The fractional length change is

$$\frac{\Delta l}{l} = \frac{f/A}{Y} \tag{4.18}$$

Here, Y is Young's modulus. For stainless steel, $Y \cong 2 \times 10^{11}\,\mathrm{N/m}^2$. This means a 1/10% stretch would be produced if $f/A \cong 2 \times 10^8\,\mathrm{N/m}^2$. Thus, the cable stretch should be insignificant. Stainless steel can also be permanently stretched, and it can eventually break. The margin of safety is not that great. Stainless will plastically stretch at roughly triple the estimated shroud force per unit area of $1.5 \times 10^8\,\mathrm{N/m}^2$, and it can break when the force per unit area is roughly six times this large. Nevertheless, theory says the situation shown in Figure 4.13 should not have occurred. Theory is sometimes wrong.

4.11 Personal Torques

When the wind is blowing, torques are as important as forces. Sailboats can move fast only when they stay upright. On small boats, the sailor supplies much of the compensating torque needed to keep a boat balanced.

No modern sailor has gained more fame or won more awards than Paul Elvstrom, and much of his fame is due to his mastery of torques. Elvstrom invented the techniques and the equipment that make "hiking" possible. The sailor in Figure 4.7 can sit out past the edge of the boat because of hiking straps, which are an Elvstrom invention. The torque is the product of force and distance, so one wants to hike out as far as possible. One also needs to maintain this stressful position for long periods of time. Elvstrom installed special hiking apparatus in his garage and practiced dry-land hiking for hours at a time. Of course, Mr. Elvstrom needed additional sailing skills to be named the "Danish Sportsman of the Century," but hiking helped.

5

SEE HOW THE MAINSAIL SETS

A sail's job depends on the relative wind direction. Downwind, the wind is perpendicular to the sail's surface, and the drag force pushes the boat along. For upwind sailing, drag is an impediment, and wind's glancing blow to the sail produces mostly lift. A large lift-to-drag ratio becomes increasingly important when sailing more toward the wind.

Birds and airplanes remain aloft because of large lift-to-drag ratios, and sailors can gain useful hints about sail trim by observing the wings of our flying friends. Since a sail is made of a single sheet of sailcloth, it can only roughly approximate the more effective airfoil shapes of relatively thick bird and airplane wings. Bat wings are thin, but the spectacular maneuverability and bug-trapping abilities of bat wings are not of primary interest for most sailors.

An ideal sail should have an adjustable shape to efficiently accomplish its many jobs. Sail flexibility allows some shape changes, and alert sailors are perpetually adjusting sail trim to improve their speeds. Because the shape alterations of a single sail are limited, many sailboats use spinnakers in addition to their regular sails. Spinnakers are useful only when the wind is more or less astern. Their rounded shapes are ideal for producing the downwind drag, and some spinnakers also have significant lift so they can be used on a reach where the wind is from the side. To further deal with changing wind conditions, some boats use jibs of varying shapes and sizes. Sail areas can also be reduced by reefing. In principle, the sailor with the largest storehouse of sails could sail the fastest. Cost, common sense, and restrictions for sailboat classes generally limit the choice of sail materials and the number and overall dimensions of sails.

Assume one has the insight to determine a "best" sail shape, despite all the complications of variable wind. One would hope that fastening sailcloth to a mold of this ideal shape would produce the ideal sail in the same way that a mold is used to produce ideal shapes for hulls, rudders, keels, and centerboards. This construction will usually fail because the cut of a sail alone does not determine its shape. The flexibility of sailcloth means the shape is changed by the pressures of the wind, and these pressures are not considerate enough to make the ideal shape a possibility.

5.1 Spinnaker

A simple downwind spinnaker that is similar to those commonly used on small sailboats is both nearly spherical and triangular. One can construct one of these spherical triangles as follows:

1. Snuggly attach sailcloth to a giant globe (diameter: several meters).
2. Cut the cloth along a path from the North Pole to the Equator along the Greenwich (England) meridian.
3. Continue the cut due west along the Equator to longitude 60° west.
4. Make the third and final cut north through the eastern tip of Nova Scotia in Canada, back to the North Pole.

The result is a reasonable approximation to a real spinnaker shape. A spinnaker modeled from a section of a giant football or an enormous egg might be even more effective and may more closely resemble the spinnakers shown in Figure 5.1.

Even though a spinnaker is built to have a roughly spherical shape, it can easily be distorted. When the wind dies, sailors are often disappointed to see their spinnakers change from pleasantly bulging efficient sails to drooping, multifolded curtains. Even when a spinnaker is filled, it can assume a wide variety of shapes. A reasonable starting point for describing sail shapes rests on the approximation of invariant Gaussian curvature. To explain this, we digress.

Figure 5.1   Twin spinnakers. (Photograph by Vicki Woods and Bruce Beckert. With permission.)

5.1.1 Gaussian Curvature

The mathematics underlying the shapes of idealized spinnakers (and sails in general) has a distinguished pedigree. When Carl Friedrich Gauss (an "A-list" mathematician if there ever was one) proved that Gaussian curvature was an isometric invariant in 1828, he was so proud of this proof he called it his *theorema egregium*, which means "remarkable theorem." This theorem is a fundamental concept of differential geometry. The basic idea can be seen in the real-world examples of sails.

A sail is a surface that can take on many shapes. An isometry is a special kind of shape change that does not stretch or shear the sail material. In other words, sailcloth fibers are not stretched and crossed fiber remain perpendicular. For isomeric changes, the size of a square piece of sail cannot be increased. Also, the square cannot be changed to a rectangle or a parallelogram. This is an idealization, since all sails stretch a little.

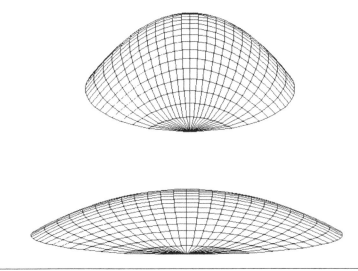

Figure 5.2 An isometric distortion of a section of a spherical shape, with stretching in one direction and compression in the other.

After you eat half a grapefruit, you are left with a shell resembling a half-sphere of radius R. The Gaussian curvature of this shell is defined to be $K = 1/R^2$. Imagine this grapefruit shell to be a miniature spinnaker. You can isometrically distort this hemispherical grapefruit-spinnaker by squeezing on opposite edges of the shell. As the bend in one direction is increased, the grapefruit shell is flattened in the other direction, roughly like the example in Figure 5.2. The distorted surface is now characterized by two different bending rates, with two different radii, R_1 and R_2. Since the grapefruit is no longer spherical, its Gaussian curvature in the more general case is defined to be

$$K = 1/(R_1 R_2) \tag{5.1}$$

Gauss's remarkable theorem tells us that the K of the distorted grapefruit is exactly the same as the K of the original spherical surface. For those who don't like grapefruit, the same trick works with a baseball cap or a yarmulke.

An egg-shaped spinnaker has different values of K at different points. The Gaussian curvature of an egg is largest at its pointed end, smaller at the rounded end, and smallest at its fattest part between the ends. At this fattest part, the two radii that determine K are associated with the path around the egg's belly and the path from end to end. Any isometric distortion of an egg-shaped spinnaker must keep the local K at each point unchanged.

5.1.2 Spinnaker Shape Changes

The invariance of Gaussian curvature gives a sailor some hints for how to adjust sail shapes. On many boats, the corners of a spinnaker are fixed to the mast at the top, to a pole on the windward side of the boat, and to a single line to the lee. For this geometry, only three coordinates are adjustable. These are the vertical and horizontal positions of the pole and the length of the line. Raising the pole is analogous to squeezing the grapefruit in the vertical direction. Invariant Gaussian curvature means the increased vertical bending must be compensated by a decreased bending in the horizontal direction, so a horizontal section of the spinnaker appears "flatter" or less curved. Similarly, pulling the line to stretch the spinnaker along its lower part (the "foot") will straighten the foot but increase bending in the vertical direction near the foot. The generalization is clear. When you bend a spinnaker (or any sail) in one direction, it becomes flatter in the other direction.

The other sails (mainsail and jib) are much flatter than the spinnaker, so their Gaussian curvatures are much smaller. Even so, the isometric invariance of Gaussian curvature can help one to understand general sail design and trim.

5.1.3 Make Your Own Sail

Consider again the insightful sailor who knows exactly what sail shape will produce the best sailing. Knowing the sail shape means knowing the Gaussian curvature at each point on the sail. This is sufficient information to construct the sail.

The geometry of curved surfaces is peculiar in many ways. For example, a circle's circumference is no longer 2π times its radius (assuming the radius is measured along the sail surface). For a disk with a small radius, the relation between the circumference, the radius, and the Gaussian curvature is

$$circumference \cong 2\pi \cdot radius \left(1 - \frac{K}{6}(radius)^2 \right) \qquad (5.2)$$

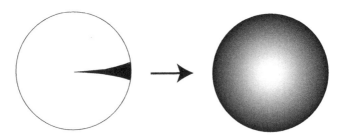

Figure 5.3 A circle with slit removed becomes a curved surface when reattached.

The term that is proportional to K is curvature's modification of plane geometry. In principle, this formula could be used to construct a sail with the desired Gaussian curvature at each point. Remove a slit of width $\pi (radius)^3 K/3$ from a circular piece of a sail, pull the crack shut, and sew the edges together, as shown in Figure 5.3.

The "doctored" patch has the desired curvature. Sew together a patchwork of these pieces, each with its own Gaussian curvature to form the sail with the desired shape.

A patchwork of circles is a tedious and expensive way to make a sail. It is far simpler to produce curved surfaces by sewing together tapered strips of sailcloth, similar to the way tapered map strips are glued to the surface of a globe. As one moves a distance d along the strips, the width shrinks at a rate determined by the Gaussian curvature.

$$w(d) = w_0 \left(1 - \frac{1}{2} K d^2 \right) \tag{5.3}$$

This sail-making technique is called "broadseaming." Material within the strips has no curvature, but averaged over distances that are large compared to the strip width W_0, the desired curvature is obtained, as shown in Figure 5.4.

5.1.4 Stress

Stress on a small piece of sail is composed of opposing forces trying to pull the sail apart. The opposing forces mean stress is not simply a vector. It is a tensor, denoted $\overleftrightarrow{\tau}$ with a double-pointed arrow decoration. A vector is specified by a direction and a magnitude. It is commonly

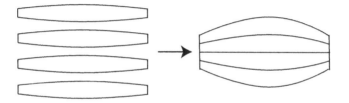

Figure 5.4   Tapered strips are needed to make the curved surface of a globe or a spinnaker.

displayed as an arrow. A tensor in the two dimensions of the sail surface is specified by a direction $\hat{n}$ and two stress magnitudes, τ_1 and τ_2, that are parallel and perpendicular to $\hat{n}$. Examples stress tensors, displayed as perpendicular pairs of double-pointed arrows, are shown in Figure 5.5.

All the stresses of Figure 5.5 occur in sails. The symmetric stress (a) approximates the center region of a spinnaker. The one-dimensional stress (b) occurs along the free edges of a sail, with the double-pointed arrow parallel to the sail edge. The anisotropic stress of (c) is the general case where the forces are larger in one direction. Finally, the compression stress shown in (d) cannot be supported by sailcloth, but battens (rigid support rods placed in the sail) can deal with compression.

Sail stresses are expressed in units of Newtons/meter. The force on a piece of sail of width, w, along a stress direction is τ_1 (or τ_2) times the width, w.

A shear force changes a square piece of cloth into a parallelogram. There are no shear forces in a coordinate system oriented along a stress tensor axis $\hat{n}$. Most sailcloth is good at resisting forces parallel to the fiber orientation, but they are less able to resist shear. Thus, it is desirable to align the sailcloth fibers along the axes of the stress tensor so that shear is not a problem. The isotropic stress shown in Figure 5.5a produces no shear in any direction.

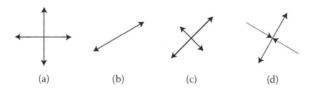

(a)   (b)   (c)   (d)

Figure 5.5   Four different stress tensors. (a) An isotropic stress. (b) Stress in only one direction. (c) Anisotropic stress. (d) A compression along one direction and tensile stress in the other direction.

When sailing downwind, the isotropic spinnaker stress is characterized by single number τ that gives the magnitude of the stress in any direction. The pressure P and the stress τ work against each other. The pressure pushes the sail forward, and the stress accompanied by bending pulls it back. Because bending is inversely proportional to the radius of curvature, the forces are balanced when

$$bending = \frac{1}{R_1} + \frac{1}{R_2} = \frac{pressure}{stress} = \frac{P}{\tau} \tag{5.4}$$

The two curvature radii are about the same for a spinnaker with a roughly spherical shape, so $R_1 \cong R_2 \cong R$. Thus,

$$\tau \cong \frac{1}{2}RP \tag{5.5}$$

In a Fresh Breeze, $P \approx 20$ N/m². A spinnaker radius of curvature is on the order of 4 m, so $\tau \approx 40$ N/m. This stress could not even tear a piece of paper. Since much of a spinnaker is subject to relatively small stresses, spinnaker material can be lighter than ordinary sailcloth. Reinforcements are needed near the spinnaker corners where stresses are concentrated.

5.2 Mainsail and Jib

Mainsails and jibs resemble nearly flat triangles. One might assume such sails are simple, but sailors see subtle complexities in the triangles that propel their boats. Even the language of sails is complicated. Just as some people know many words to describe camel anatomy or the different types of snow, sailors have their own vocabulary for the parts of a sail, which are labeled in Figure 5.6a.

The sail shape in Figure 5.6 is a simplification. Most sails have additional area because the leech is curved. Battens (horizontal plastic strips embedded in the sail) allow the sail to extend beyond the line between the head and the clew. The extra area is called the "roach." Full-length battens that extend from the luff to the leech, combined with high-tech sail materials that stretch very little, can produce a large roach and more effective sail shapes.

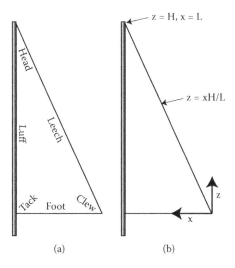

Figure 5.6   (a) Parts of a sail. The same nomenclature applies to the jib and even the spinnaker. (b) The coordinates used to describe points on the sail's surface.

Many sailboats use two sails upwind—a mainsail and a jib (sometimes called headsail or genoa if it is big). The luff of the mainsail is attached to the mast. The jib's luff is attached to the forestay, which is a cable running from the mast to the bow. The forestay is one of three or more cables used to keep the mast upright. The jib and mainsail each deflect the wind resulting in mutual interaction. The net effect of this interaction is not as large as one might expect. There is generally little difference in the speeds of two similar sailboats when one is rigged with a single sail and the other uses both a mainsail and a jib. Of course, the total sail areas must be the same for a fair comparison.

If a sailor is unusually imaginative and even more unusually prosperous, other sail designs, like the one in Figure 5.7, can be tried.

The nonrigid nature of sailcloth, combined with the mysteries of the wind's pressure variations, make it extremely difficult to analytically characterize real sail shapes. Sailmaking is an art too subtle for me to describe. However, some elementary ideas about sails are described here by considering the terribly oversimplified example of sails characterized by two approximations.

- The Gaussian curvature is zero everywhere on the sail.
- The sail's surface is subject to a uniform pressure.

Figure 5.7   Nonstandard sails on a nonstandard sailboat. (Photograph by Debbie Kennedy. With permission.)

Modifications of sail shapes that come about from Gaussian curvature, nonuniform pressure, and other complications are discussed briefly in Section 5.3.

Vanishing Gaussian curvature does not mean the sail surface is flat like a board. Instead, each point on the surface must lie on a straight line. Simple experiments with a piece of paper cut to the shape of a miniature sail show this. No matter how you (gently) bend the paper sail, perfectly straight lines traverse the model sail from one edge to the other.

Assume a sail has the simple geometry shown in Figure 5.6. It is a right triangle, with a luff of height H attached to the mast. The foot of length L is attached only at the corner (the clew). The assumptions of uniform pressure and vanishing Gaussian curvature mean the sail shape is entirely determined by the position of the clew. If the clew is stretched out and down as far as possible, the sail will be perfectly flat. In practice, this flat shape could be achieved only on calm days. A wind bulges the sail to one side and gives it a rounded shape, which is a good thing because a perfectly flat sail is never desirable. Sailors adjust the position of the clew to vary the curvature and shape of a sail. To illustrate the flexibility of this flat sail, three examples are described. The first example keeps the leech tight but reduces the tension on the foot. The second is characterized by a tight foot with reduced tension on the leech. The final example is a compromise in which both foot and leech tensions are reduced in such a way that the sail shape is particularly simple. Calling this last example "Perfect Blend" is a value judgment.

5.2.1 Tight Leech

If the clew is moved slightly toward the mast while the leech is kept tight, the sail will be curved and the foot will have the shape of a circular arc (roughly a parabola). The sail shape is derived by minimizing the potential energy of the sail subjected to a uniform pressure.

Since the sail has vanishing Gaussian curvature, every point on the sail must lie on one of an array of straight lines that radiates from the head to the foot. This geometry gives the sail displacement shown in Figure 5.8. Equations describing the sail shapes are given in Section 5.2.4.

5.2.2 Tight Foot

The Tight Foot sail shape is quite different from the Tight Leech. The difference is produced by pulling the clew tightly along the boom and allowing it to move up slightly along the leech. Now the parabolic shape is along the leech, and the straight lines radiate from the tack to the leech. Minimizing the energy gives the sail shape shown in Figure 5.9, which resembles the shape in Figure 5.10. The equation describing this sail is also given in Section 5.2.4.

5.2.3 Perfect Blend

A more sensible sail shape is obtained when a moderate force is applied along both the foot and leech of the sail. This shape is particularly simple because the force on the clew (equivalently, the clew position) is chosen to make the straight lines on the sail's surface vertical and parallel to the mast, as is shown in Figure 5.11 and in the sail of Figure 5.12, which resembles this shape. The Perfect Blend sail shape

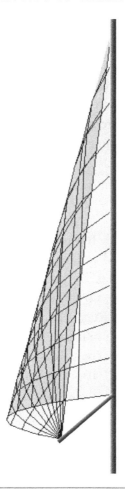

Figure 5.8 A Tight Leech sail with vanishing Gaussian curvature subjected to a uniform pressure. The maximum sail displacement has been exaggerated to show the shape.

is described in more detail in Section 5.2.4 because it is a fairly realistic representation of a practical sail shape.

The Perfect Blend sail is a reasonable model for the shape and forces one would expect from a typical sail. Some of its properties are discussed here.

The maximum displacement of the Perfect Blend sail does not occur at one point. Instead, there is a vertical line of "maximum draft" that is about 42% of the distance from the luff to the clew. The sail shapes at the foot for large, medium, and small forces applied to the clew are shown in Figure 5.13. These shapes have a reasonable resemblance to an airfoil that our intuition tells us is a desirable property of an efficient sail.

Figure 5.9   The Tight Foot sail with vanishing Gaussian curvature subjected to a uniform pressure. Again, the maximum displacement has been exaggerated.

The maximum displacement, y, and the horizontal clew force, F_x, are inversely related. For the Perfect Blend shape,

$$\frac{y(\text{max})}{L} = \frac{2}{9\sqrt{3}} \frac{A \cdot P}{F_x} \tag{5.6}$$

One can estimate the horizontal clew forces, F_x, for a Thistle sailboat in a Fresh Breeze, assuming the total force on the sails is $A \cdot P \cong 370$ N.

Figure 5.10 A sail resembling the Tight Foot shape. (Photograph by Sally Snowden. With permission.)

If $y(\text{max})/L$ is roughly 10% and if the jib is half as big as the mainsail, one obtains

$$F_x(mainsail) \cong 320 \text{ N}$$

$$F_x(jib) \cong 160 \text{ N}$$

(5.7)

The mainsail force that holds the clew away from the mast is supplied by fastening the sail to the boom, so sailors do not need to keep pulling the sail away from the mast to maintain its shape.

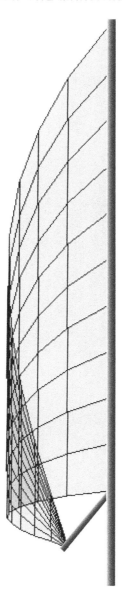

Figure 5.11   The Perfect Blend sail. The displacement of the sail does not change with height.

The force, $F_x(jib)$, must be added to a comparable or larger vertical force to maintain leech tension. The resulting total force in a Fresh Breeze needed to trim the jib is comparable to the force needed to lift a fairly large dog. The forces scale with the sail area, so winches, pulleys, and strong arms are sometimes needed for jib trim on larger sailboats.

Figure 5.12   A sail resembling the Perfect Blend shape. (Photograph by Sally Snowden. With permission.)

A special and unusual property of the Perfect Blend shape is the simplicity of the stress tensor. The results in Section 5.2.4 can be used to show that all the stresses are along lines from the clew to the luff. There are no transverse stresses, so the stress at every point in the sail can be represented as a single arrow with double points (not two

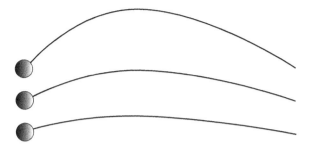

Figure 5.13 The Perfect Blend sail shape along the foot. From top to bottom, *y*(max)/*L* is 1/5, 1/10, and 1/20. The curvature and draft are inversely proportional to the tension. The circles on the left represent the masts.

double-pointed arrows), as shown in Figure 5.5b. The sail stresses are illustrated in Figure 5.14.

The magnitude of the stress in the Perfect Blend sail shape is conveniently written in polar (r, ϕ) coordinates. Here, r is the distance from the clew where the force is applied, and ϕ is the angle above horizontal. Thus, the foot of the sail lies on the line corresponding to $\phi = 0$ and $0 \le r \le L$. In these coordinates, the magnitude of the stress radiating from the clew is

$$\tau(r,\phi) = F_x \frac{1}{r} \frac{L}{H} \frac{1}{\cos^3(\phi)} \tag{5.8}$$

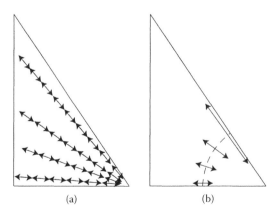

(a)   (b)

Figure 5.14 (a) Arrows indicate the direction of stress for the Perfect Blend sail. Crowded arrows near the clew (lower right) are indicative of larger stress. (b) The lengths of the arrows are proportional to the magnitude of the stress at a fixed distance from the clew. This shows the stress is largest along the leech.

Some of Equation 5.8 is intuitive. It is reasonable for the stress to be proportional to the applied force. The factor of $1/r$ means the stress is concentrated at the clew. In practice, sails are reinforced near the clew to withstand this stress and limit sail deformation. The $1/\cos^3(\phi)$ term in Equation 5.8 is more surprising. It means that the stress along the foot is typically ten times smaller than the stress along the leech. This has important consequences when considering sail strength. For a traditional woven sail, the strongest fibers (called "fill") should be parallel to the leech where the stress is largest. The perpendicular fibers (walled warp) bend around the fill, so they stretch more easily. For high-tech sails, the considerations are similar, but the stress resistance may be provided by carbon fibers.

Summing all the force vectors associated with the stress gives the total force of the sail on the clew. The result of this summation is fairly simple. One finds that the total force for the Perfect Blend sail is directed toward a point halfway up the mast. Typically, this means the vertical force on the clew is comparable to or larger than the horizontal force. The total force is thus typically 50% larger than the horizontal force. For the example of the Thistle sailboat in a Fresh Breeze, the force needed to maintain the Perfect Blend shape for a jib with a y(max) about a tenth of L is roughly 250 N.

5.2.4 Sail Shape Equations

The assumptions of uniform pressure and vanishing Gaussian curvature is sufficient to allow one to derive the formulas for the sail shapes shown in Figures 5.8, 5.9, and 5.11. Although formulas for all three shapes are presented, a sketch of the derivation is given only for the Perfect Blend sail.

First, some coordinates are needed, as illustrated in Figure 5.6b. The flat sail lies in the x-z plane, which means $y = 0$. The pressure of the wind bulges the sail out a bit in the y-direction. The resulting sail shape is specified by the displacement $y(x, z)$ for each point on the sail. The clew is placed at the origin of the coordinate system, $x = 0$, $y = 0$, and $z = 0$. The foot extends a distance L along the x-axis to the mast whose base is at $x = L$, $y = 0$, and $z = 0$. The mast is attached to the sail's luff and extends to the sail height $z = H$. That means the sail's leech lies on the line $z = x \cdot H/L$. (The drawing in Figure 5.6b is a little unorthodox because x increases from right to left.)

The Tight Leech sail is described by

$$y(x, z) = 4 \cdot y(\text{max}) \frac{(Hx - Lz)(L - x)}{L^2 (H - z)} \qquad (5.9)$$

Here, $y(\text{max})$ is the maximum displacement that occurs halfway along the foot at $x = L/2$ and $z = 0$. The ratio $y(\text{max})/L$ character-izes sail shape. Pulling the clew out enough to make $y(\text{max})/L < 10$ gives the sail a "flat" appearance. Moving the clew toward the mast makes a "full" or "baggy" sail with a larger value of $y(\text{max})/L$.

The Tight Foot sail is described by

$$y(x, z) = 4 \cdot y(\text{max}) \frac{(1 - x/L)(z/H)}{(1 - x/L) + (z/H)} \qquad (5.10)$$

For this case, the maximum displacement occurs halfway up the leech at $x = L/2$, $z = H/2$. As with the first example, $y(\text{max})$ is determined by the force applied at the clew.

Because the straight lines are vertical, the Perfect Blend sail displacement does not depend on z, so $y(x, z) \rightarrow y(x)$, and $y(x)$ describes the sail shape for all heights z between the foot and the leech directly above it. This simplifies the algebra needed to obtain the shape. The Perfect Blend sail shape is also determined by the force $\vec{F}$ applied to the clew. This force produces a stress tensor $\overleftrightarrow{\tau}(x, y)$ in the sail, which in turn determines the horizontal ten-sion $T(x)$. (Formally, the stress tensor $\overleftrightarrow{\tau}$ described in Section 5.1.4 has the form of a 2×2 matrix and $T(x)$ is the (x, x) component of this matrix.) For each x, the tension is uniformly distributed over the vertical distance from foot to leech. Multiplying the tension $T(x)$ by the sail height at x, which is $h(x)$, gives the x-component of the force applied to the clew, F_x. Because $h(x) = xH/L$, this means

$$T(x) = \frac{1}{x} \frac{L}{H} F_x \qquad (5.11)$$

The $1/x$ in this equation means the tension is largest near the clew at $x = 0$, where it is supported by only a small amount of sail, and it is smallest at the mast.

The pressure P and $T(x)$ work against each other, with P increasing the bending and $T(x)$ making the sail flatter. The balancing of these two effects means

$$bending = \frac{pressure}{tension} = \frac{P}{T(x)} \qquad (5.12)$$

The bending is essentially the second derivative of $y(x)$, so using the expression for $T(x)$ gives a differential equation for this Perfect Blend shape

$$\frac{F_x}{x}\frac{L}{H}\frac{d^2y}{dx^2} = -P \qquad (5.13)$$

The solution is

$$\frac{y(x)}{L} = \frac{1}{3}\frac{A \cdot P}{F_x}\left(\frac{x}{L} - \left(\frac{x}{L}\right)^3\right) \qquad (5.14)$$

Here, $A \cdot P/3$ (one-third of the sail area times the pressure) is the same as the y-component of the sail's force on the clew. This shape represented by Equation 5.14 is shown in Figure 5.13 for three different values of the applied force F_x.

5.2.5 Sail Characterization

A single sail with vanishing Gaussian curvature can assume a variety of shapes. The Tight Leech, Tight Foot, and Perfect Blend are just three examples. A variety of numbers simplify the characterization at sail shapes. These numbers are compared here for the three example sails.

The angle θ between the boom and the line tangent to the luff is the entry angle. Vanishing Gaussian curvature means θ is the same at every point on the mast for all three example sail shapes.

For the Perfect Blend shape shown in Figure 5.15,

$$\theta = \frac{180}{\pi}3\sqrt{3}\frac{y(\text{max})}{L} \qquad (5.15)$$

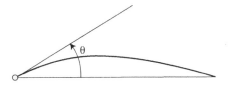

Figure 5.15   The entry angle θ is between the luff and the horizontal boom. The example shown here is the Perfect Blend sail along the foot.

For the reasonable case where $y(\max)/L \cong 1/10$, the entry angle is about 30°. For the other two example sail shapes, the Tight Leech and the Tight Foot, the entry angle is described by a similar expression. The coefficient of $y(\max)/L$ is slightly different for the three sail shapes. For all three sail examples, entry angle is proportional to the maximum draft, which means it is inversely proportional to the force applied at the clew.

For each height on the sail, its cross section is a simple curve. Two cross sections are shown for the three example sail types. The longer curve displays the shape at the foot, and the shorter shows the sail shape halfway up the mast. For each curve, the straight line from the luff to the leech is the "chord."

Three numbers are used to describe these curves. The twist angle is the angle the chord is rotated as one moves up the sail. The camber ratio is the largest draft d divided by the chord length. The position of maximum draft is expressed as a fraction of the chord length. For the three example sail shapes, these quantities can be calculated for any height using Equations 5.9, 5.10, and 5.14, but Figure 5.16 shows the basic trend.

5.2.5.1 Twist   The Tight Leech sail has no twist. The Tight Foot sail has the most twist, and the Prefect Blend lies between the two extremes.

5.2.5.2 Camber Ratio   The camber ratio of the Tight Leech sail does not vary with height. This ratio increases with height for the Tight Foot example and decreases with height for the Perfect Blend sail.

5.2.5.3 Maximum Draft Position   The maximum draft position is always midway between the luff and the leech for the Tight Leech example. For the Perfect Blend sail, the maximum draft at the foot is ~42% back at the foot and moves so it is centered near the head.

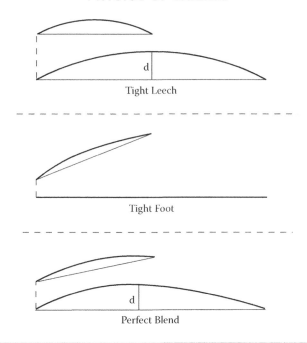

Figure 5.16 Cross sections of the three example sails at the foot and halfway up the mast.

For the Tight Foot example, the maximum draft is near the luff at the bottom of the sail. It also moves to halfway back at the head.

5.2.6 Applying the Forces

In the crudest of nonnautical terms, "the boat won't go until someone pulls on a rope." Stated more formally, a sail's shape and position is determined by the magnitude and direction of the applied force. It is the sailor's job to apply the force properly.

5.2.6.1 Sail Shape   Sail trim is relatively simple for the example jib in Figure 5.17. The vector showing the sail force is directed toward the center of the luff. This is roughly the force direction associated with the Perfect Blend sail shape. An equal and opposite "sheet force" must be supplied by the sailor pulling on a "jibsheet." (It is bad form for a sailor to call a rope a "rope." It's a "sheet," a "line," or a "halyard," depending on its use.) Since the force can be more than a couple of hundred Newtons for even a small jib, some of mechanical advantage (e.g., a winch) is often needed to properly trim the sail. If the direction

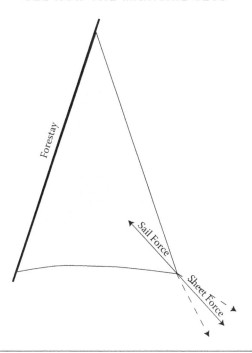

Figure 5.17   The force applied to the jib sheet is equal and apposite the force of the sail. The dotted lines denote alternative force directions that would change the sail shape.

of the applied force is changed, as indicated by the dotted alternative sheet forces, extra tension will be applied to the leech or the foot. This will change the sail shape to more closely resemble the Tight Leech and Tight Foot examples. In practice, the jib sheet runs through a "block" (pulley) on the deck. Moving the block back tightens the foot, and moving the block forward tightens the leech.

The force applied to the mainsail is essentially the same, but the force is distributed in little pieces, as illustrated in Figure 5.18. The clew is attached to the boom, so the sailor is not required to continually pull out on the mainsail. The position of the clew along the boom is adjustable with an "outhaul" that delivers the horizontal component of the sail force. [The vertical component of the sail force is also supplied by the boom, and the boom pulls down on the sail because it is attached to a mainsheet and a vang.] The boom is free to rotate about its fixed pivot point attached to the mast. The force applied to the sail is determined by the condition that the sum of the torques applied to the boom must vanish. As described in Chapter 4, each

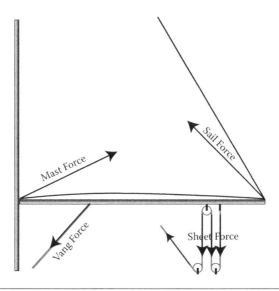

Figure 5.18 The forces applied to the boom.

torque is the product of a vertical force multiplied by a horizontal distance. Applying the torque rule to the example shown in Figure 5.18 means the vertical component of the sail force is roughly 80% of the total sheet force plus 30% of the vertical component of the vang force. Because the mainsheet makes three trips to the boom, the total sheet force is three times the force the sailor must apply to the end of the mainsheet.

The boom is subjected to considerable compression as well as bending forces, so it must be quite strong. The mast force on the boom is determined by the condition that the sum of all forces applied to the boom must vanish. When the vang tension is very large, some booms will bend, as shown in Figure 4.7.

5.2.6.2 Sail Position This focus on sail shape has ignored the most important angle for sailing. The sail must be properly oriented with respect to the wind. This means the sail must be trimmed close to the center of the boat for upwind sailing, but when sailing downwind or on a reach, the boom must be allowed to swing out. Releasing tension on the mainsheet allows the sail to move out for downwind sailing. This presents a problem because the vertical component of the mainsheet force nearly vanishes. Without the vertical force supplied by the vang, the sail could rise up in heavy winds and become less efficient.

Many sailboats have adjustments that can move the mainsheet trim points to the edge of the boat. This allows the mainsheet to be more nearly vertical. The additional vertical force is especially helpful when the wind is strong.

5.3 Real Sails

Much has been ignored in this characterization of sails. Some obvious oversights are

1. The pressure on a sail's surface is not uniform.
2. Sailcloth stretches.
3. Sails are constructed with nonzero Gaussian curvature.
4. The mast bends on many smaller sailboats.
5. Luff tension can be adjusted.
6. The foot of many sails, especially mainsails, is attached to the boom.
7. Battens (solid strips extending to the leech) are placed in the sail.

Comments on some of these oversights follow.

5.3.1 Pressure Variation

On real mainsails, the pressure is typically largest near the mast (but not too near the mast), and there is also a smaller viscous force from the wind pulling along the sail's surface. A more realistic description of the forces results in modest (but possibly important) changes in the calculated sail shapes. The task of predicting and controlling sail shape under real sailing conditions is daunting. The problem is circular.

a. The air pressure and flow determines the sail shape.
b. The sail shape affects the airflow.
c. But the airflow determines the pressure, so you return to Step a without a solution.

An example of nonuniform pressure is considered for the Perfect Blend sail shape. When the pressure depends on the position and

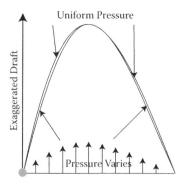

Figure 5.19 A comparison of sail shapes when the pressure is uniform (wider curve) and when the pressure is largest near the center of the sail (narrow curve). The arrows at the bottom indicate the assumed form of the pressure variation. The curves are scaled to coincide at maximum draft.

$P \rightarrow P(x)$, the generalization of Equation 5.13 is

$$bending \propto x \cdot P(x) \tag{5.16}$$

Two sail shapes with different pressure distributions are compared in Figure 5.19. The draft is exaggerated so the difference in shapes can be seen. The slightly wider "uniform pressure" curve for the foot of the Perfect Blend sail is the same as Figure 5.13, but with a magnified vertical scale. The narrowed "pressure varies" curve is produced by the pressure distribution shown by the arrows at the bottom of the figure.

The result shown in Figure 5.19 for the Perfect Blend example is fairly general. The nonuniform pressure distribution has a small effect on most sail shapes.

5.3.2 Stretching, Bending, and Other Complications

5.3.2.1 Stretching   Stress applied to a sail can change its shape. When a piece of sail is pulled in a direction parallel to its fibers, its length d can increase to $d + \Delta d$, with the extra length given by

$$\frac{\Delta d}{d} \approx \frac{stress}{thickness \cdot Y} \tag{5.17}$$

An example illustrates the application of this formula. Assume one has a long piece of sailcloth that is 5 m long and 1/2 m wide.

Two people grab opposite ends of this cloth and pull with opposing forces of 50 N. The "stress" is this force divided by the sailcloth width, so *stress* = 100 N/m. The stress must be divided by the thickness of the sailcloth to obtain the fractional change in length. Typically, the thickness is given indirectly in terms of the mass of a given area of sailcloth, and it is often expressed in extraordinarily obscure units. Assume the sailcloth is a spinnaker made of relatively stretchy nylon that is only 3/10 mm thick. The second term in the denominator, Y, is Young's modulus, which quantifies a material's ability to resist stretching. For nylon, $y \cong 3 \times 10^6$ N/m. Substituting these estimates into Equation 5.17 gives $\Delta d/d \approx 10\%$, which means the 5 m piece of light spinnaker would stretch half a meter. The Fresh Breeze estimate in Section 5.1.4 for the center of the spinnaker was 40 N/m, so a typical $\Delta d/d$ in typical sailing conditions is significant, but usually less than 5%.

As one might expect, there is more to stretch than the single formula for $\Delta d/d$ suggests. Young's modulus often differs for the two different fiber directions in the sailcloth, and there is yet another coefficient for the effect of stretching a sail along the diagonal between the fiber directions. Diagonal stretching corresponds to a shear of the material. It is easier to shear sailcloth than to stretch it along the fibers. These generalizations mean results obtained from Equation 6.17 are only first approximations.

Sail stretch is not always a bad thing. A spinnaker's elasticity can improve stability because abrupt motion of the boat can be absorbed through stretching rather than transmitted across the sail. Limited stretching does little to alter the simple rounded shape of a spinnaker.

Stretch is more troublesome for mainsails and jibs. The stresses are much larger for these sails, and the stretching can significantly change the shape of these more nearly flat sails. For sailors of modest means, mainsails and jibs are typically made from polyester (traditionally Dupont's Dacron), which is about one-third as stretchy as nylon. Also, mainsails and jibs are quite a bit thicker than spinnakers. This results in manageable, but not negligible, stretching. Sailors of immodest means can purchase high-tech sails that use aramids (e.g., Kevlar) and/or carbon fibers. These materials can be 10 times more resistant to stretching than Dacron. As time passes, they are becoming less expensive.

When a sail is stretched, its shape is deformed. For the Perfect Blend sail shape, the largest stress (and stretch) is along the leech, as illustrated in Figure 5.14b. The stretched leech will allow the maximum displacement $y(\text{max})$ to increase. This is especially noticeable when a sail is trimmed to be fairly flat. If $\Delta y(\text{max})$ is the change in $y(\text{max})$ caused by sail stretching,

$$\frac{\Delta y(\text{max})}{y(\text{max})} \approx \left(\frac{L}{y(\text{max})}\right)^2 \frac{\Delta d}{d} \qquad (5.18)$$

Since $L/y(\text{max})$ is on the order of 10, a change in the length of the leech of only one part in 1,000 can cause a significant 10% change in the maximum value of y along the leech. Some of the stretch can be corrected by applying additional vertical force to the clew. However, this compensation has its limitations. When the edge of a sail is stretched, the circumference of a circle drawn on the sail can increase even if its radius remains constant. This corresponds (through Equation 5.2) to a decrease in the Gaussian curvature, which cannot be restored by altered sail trim. Also, since the stretching varies with the wind's pressure, one cannot satisfactorily compensate for this effect in both light and heavy winds.

Over time, some sail stretching becomes permanent. A sail's shape does not last forever. Sailors interested in maximum speed buy new sails much more frequently than they buy new boats. Boat stretching is less of a problem.

5.3.2.2 Gaussian Curvature   The Tight Leech, Tight Foot, and Perfect Blend sail shapes with vanishing Gaussian curvature are only simplifications. Built-in Gaussian curvature significantly modifies previous comments. As a very rough approximation, built-in Gaussian curvature adds to the draft in the center of the flat sail shapes. A sail's built-in curvature limits the sailor's ability to flatten a sail. As an extreme example, no amount of stress applied to a spinnaker can make it flat.

In many cases, the mainsail is attached along the length of the boom, which necessitates significant Gaussian curvature near the foot. Otherwise, the sail would resemble the Tight Foot sail shape. Extra sailcloth inserted near the foot allows the sail to have a more reasonable shape. The extra cloth allows horizontal curvature just a short distance above the foot.

Sail stretch and Gaussian curvature combine to make sail trim much more difficult to understand. As an example, assume a sailor pulls down harder on the boom. This will straighten the leech. However, if the Gaussian curvature is nonzero, the straighter leech means more curvature in the direction perpendicular to the leech. The sail may become "cupped" at its back edge, as in the grapefruit spinnaker example of Figure 5.2. However, if the additional stress on the leech is sufficient stretch the sail and to decrease the Gaussian curvature, the result could be a flatter sail instead of a sail with a cupped leech.

5.3.2.3 Bending Masts The masts on many smaller sailboats are flexible. They can be bent aft either by tension applied to the leech or a backstay. Sometimes the boom can also bend, as can be seen in Figure 4.7. Mast bending can flatten a sail, provided the sailcloth allows the small amount of distortion needed to accommodate the shape change. In general, the change in sail shape produced by mast bending is not an isometry, so the Gaussian curvature cannot be strictly invariant. A high-tech sail with almost no stretch is not compatible with a bending mast.

The mast bending effect can be seen by attaching strings to a sail that has a "full" center section. This full sail has nonzero draft even when the edges of the sail are pulled so tight that they are straight. The fullness of the sail is shown in Figure 5.20a by strings that run from the clew to the luff. The string curvature represents the draft of the sail. The bottom string of length L is straight. It runs along the foot from clew to tack. The top string is also straight. Its length is $\sqrt{L^2 + H^2}$, and it runs along the leech from clew to head. The other strings are not straight because they are attached to the sail with some draft in the middle. For example, the center string runs from the middle of the luff to the clew. This string is a little longer than plane geometry requires because the sail is bulging out in the middle. If the middle string has length $(1+\varepsilon)\sqrt{L^2 + (H/2)^2}$, the ε characterizes the extra length, which means the draft in the center of the sail is roughly

$$\frac{y(\text{max})}{L} \approx \sqrt{\varepsilon} \tag{5.19}$$

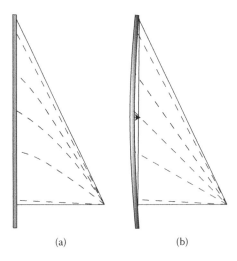

(a)   (b)

Figure 5.20   (a) The dotted lines are straight along foot and leech but bent in between because the sail has a built-in fullness. (b) Bending the mast eliminates the fullness and straightens all the dotted lines.

This draft is present even when the sail is trimmed so tightly that the foot and leech are straight.

In principle, $y(\text{max})$ can be reduced to nearly zero by bending the mast. The bent mast is shown in Figure 5.20b along with a straight line drawn from the sail head to the tack. The bending is characterized by the distance d between the mast and the line, as indicated by the little arrow in Figure 5.20b. If $d \approx \varepsilon H$, where ε characterized the extra length of the center string, the strings should all be pulled straight. This means the bending needed to flatten a sail is roughly

$$\frac{d}{H} \approx \left(\frac{y(\text{max})}{L} \right)^2 \qquad (5.20)$$

Typically, $y(\text{max})/L$ for the straight mast is less than 0.2. Thus, only modest mast bending with $d/H \approx 1\% \rightarrow 5\%$ should be sufficient to flatten a sail. Mast bends of this magnitude are common. One's eye is quite sensitive to bending, and a quick visual check of bending can easily overestimate its magnitude. The bending d/H is about 2% and 3% in Figures 5.20b and 5.21.

Figure 5.21   Mast bending on a Laser sailboat. (Picture by Sally Snowden. With permission.)

Mast bending is a practical method of sail control only if the sail can stretch a little. To see this, assume a sail only allows isometric deformations. If a bent mast could make the sail perfectly flat, then the sail would not fit on the mast when it was straight. You can see this by cutting a piece of paper in the shape of a flat sail with a curved mast. There is no way you can (gently) bend this paper to make the curved mast line straight. A sail that cannot easily accommodate this deformation typically develops wrinkles in its attempt to assume an unnatural shape.

5.3.2.4 Luff Tension   If a sail could not stretch, no adjustment of the distance between head and tack would be possible. When stretching is possible, changing the length of the luff also changes the sail shape. Additional luff tension is achieved with a "Cunningham." (Griggs

Swift Cunningham II won the America's Cup in 1958.) The result of increasing luff tension depends on the sail design. Because the largest sail stress is generally along the leech, many sails are aligned with fibers parallel to the leech. This means the luff is angled with respect to the sail fibers, so tension on the luff pulls diagonally on the fibers. Sails are easily distorted by diagonal tension. To see this, pull along diagonal points of any light piece of cloth. The material stretches easily along the diagonal, and cloth near the stretch line is pulled in, creating wrinkles. An analogous change can happen in a sail. The extra material pulled toward the luff can increase the attack angle and give the sail a larger displacement $y(x, z)$ near the luff. Sailors often say increased luff tension pulls the region of maximum draft forward. This is only part of a more complicated shape change that is produced by this adjustment. No attempt to present a quantitative description of luff tension is presented here because I don't understand it.

5.4 What Really Counts

Some aspects of sail trim are much more important than others. An ordering of priorities helps to keep things in perspective. They are listed here from most to least important.

- The sail force is proportional to the sail area. This nearly obvious fact tells most of the story. If a jib fails so that sail area is significantly decreased, there are no tricks that can make up for the lack of sail area.
- A sail works efficiently only if it is properly oriented with respect to the wind. Downwind, the sail orientation is nearly obvious. When sailing upwind, sailors pull the sails in so they do not luff, but do not pull them in further. On a reach, sailing roughly perpendicular to the wind, the proper sail angle is less obvious. It is tempting to pull sails in so they intersect more wind. This is a mistake because lift is mysterious and does not agree with our intuitive views.
- Sail shape also makes a difference, but the sail area and trim angle remain the first priorities. For example, if the boom is allowed to rise up when the wind increases, the sail area is effectively decreased.

Even fine refinements of shape receive a great deal of attention because there are no simple rules. There is plenty of room for controversy and experiment. Sailors interested in racing are usually subjected to strict limitations on the sail area, and they quickly learn the proper sail orientation. The most important variable left that might give an advantage is the sail shape and the adjustments that can perfect this shape.

- Surface smoothness makes an even smaller difference. Wind exerts a small tangential force on the sail as it skims along the surface. The force, which is related to the viscosity of the air (Chapter 6), is generally undesirable. The viscous force is probably smaller if the sail is very smooth. Perhaps more importantly, the sail pulls back on the wind with a force equal to and opposite the viscous force. A rougher sail can disturb the airflow pattern and decrease sail efficiency.

In principle, one should be able to deduce proper sail shape and adjustment without ever having to sail. In practice, no theory comes close to revealing all the devious complications of sails. When it comes to constructing and using sails, only practice makes perfect.

6

FLUID DYNAMICS

Sailors would like to know everything about the wind and water flowing past their boats. But the wind and water are worthy opponents of unlimited complexity. In principle, the Navier–Stokes equation explains all. Its solution could determine the lift and drag on any sail and any sailboat. Sadly, solutions cannot be found without the aid of a computer named Deep Thought. And if answers were provided, they would be difficult to interpret. So the Navier–Stokes equation must be augmented with approximations, experiments, imagination, and experience.

One might wonder why bother with the insoluble Navier–Stokes equation. It is presented here partly as a cultural diversion. It also motivates a careful look at two concepts important to sailors; viscosity and the Reynolds number. Do not be concerned if you cannot understand and solve the Navier–Stokes equation. No one can.

6.1 Navier–Stokes Equation

Mating Newton's laws with physical insight gave birth to the Navier–Stokes equation. One starts with the briefest possible summary of classical physics.

$$m\vec{a} = \vec{F} \qquad (6.1)$$

For fluids, the acceleration $\vec{a}$ is that of a small region of the fluid (air or water). The mass m becomes the fluid density ρ (mass/volume) multiplied by the small volume of fluid. There are two forces; one is produced by pressure changes and the other is associated with friction or viscosity. The viscous force appears when regions of fluid slide past each other.

137

A bare-bones version of the Navier–Stokes equation, written to resemble Newton's laws, is

$$\rho \frac{D\vec{u}}{Dt} = -\vec{\nabla} p + [\rho \zeta] \nabla^2 \vec{u} \qquad (6.2)$$

In this equation, $\vec{u}$ is the fluid velocity at a particular place and time, and $D\vec{u}/Dt$ is the acceleration of the fluid. The two terms on the right are the pressure and friction forces. The pressure force is written as $(-\vec{\nabla} p)$ because the gradient $\vec{\nabla}$ identifies the direction of increasing pressure. The ∇^2 in the viscous term $[\rho v]\nabla^2 \vec{u}$ identifies regions of fluid that are flowing slower then their averaged surroundings. Viscosity drags slower fluid ahead and puts the brakes on fluid flowing faster than its surroundings. The "dynamic viscosity" $[\rho \zeta]$ determines the magnitude of the viscous force.

The acceleration term in the Navier–Stokes equation is complicated because it is an acceleration that follows the moving fluid. That means

$$\frac{D\vec{u}}{dt} = \frac{\partial \vec{u}}{\partial t} + \left(\vec{u} \cdot \vec{\nabla} \right) \vec{u} \qquad (6.3)$$

The first term on the right of Equation 6.3 is the acceleration at a fixed point. The second term is essential because fluid can be accelerated even if the velocity at one place does not change. For example, if fluid is forced through a narrow channel as is shown in Figure 6.11, it must speed up as it enters the channel. The acceleration is proportional to both the speed at which the fluid enters the constriction and how quickly the speed changes with position. Because $\vec{u}$ appears twice in this second term, the Navier–Stokes equation is nonlinear and (generally) insolvable.

A sailor's major concern is the fluid flow and its acceleration, not the forces. Writing Newton's laws as $\vec{a} = \vec{F}/m$ suggests an equivalent Navier–Stokes equation, obtained by dividing by the density ρ.

$$\frac{D\vec{u}}{Dt} = -\frac{1}{\rho} \vec{\nabla} p + \zeta \nabla^2 \vec{u} \qquad (6.4)$$

The "kinematic viscosity" ζ on the right of Equation 6.4 differs from the dynamic viscosity through the division by the fluid density ρ. The *dynamic* viscosity [$\rho\zeta$] determines the viscous force on surfaces but the *kinematic* viscosity ζ is the fundamental fluid property that determines the flow.

The Navier–Stokes equation is incomplete without two further conditions. The first is the "no-slip condition" or the "sticky-surface condition." This requires the fluid velocity to vanish at surfaces, such as boat hulls and sails. The sticky surface condition leads to considerable confusion. If the fluid velocity vanishes at surfaces, why does it make any difference if surfaces are smooth? This is a surprisingly difficult question because it relates to the properties of boundary layers that are discussed in Section 6.4.

The second condition is the conservation of fluid. This condition is helpful because the chore of solving the Navier–Stokes equation for both the pressure p and the velocity $\vec{u}$ may appear to be hopeless. In practice, fluid conservation allows one to obtain the pressure "for free."

6.2 Viscosity

Fluid viscosity ([$\rho\zeta$] or ζ) appears in the two anchors of fluid mechanics, the Navier–Stokes equation and the Reynolds number described in Section 6.3. One cannot explain lift, drag, damping, turbulence, boundary layers, or a host of other fluid phenomena without invoking viscosity. Viscosity has a precise definition that allows one to calculate viscous forces for some especially simple geometries. First, it is important to distinguish the different types of forces and their names.

6.2.1 Viscosity and Pressure, Lift and Drag

Pressure is directed perpendicular to a surface, and the viscous force is parallel to the surface. You can recognize the difference by placing your hand out the window of a speeding car. Aligned for minimum drag, the sideways pull you feel on the back of your hand is the viscous force. Tilt your hand for maximum drag and the push on your palm derives from pressure.

Both pressure and viscosity can produce lift and drag, as is illustrated in Figure 6.1. For sailors, the wind's drag is parallel to the apparent wind direction $\vec{V}$ and the water's drag is antiparallel to

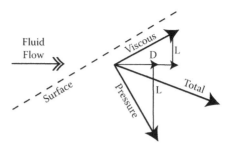

Figure 6.1   The force on a surface has a pressure component perpendicular to the surface and a viscous component parallel to the surface. Each of these forces has both lift and drag components that are perpendicular and parallel to the fluid flow. The total force is the vector sum of viscous and pressure contributions. The fluid flow vector denotes the flow a long distance from the surface.

the boat velocity $\vec{U}$. The fluid flow arrow in Figure 6.1 indicates the direction of fluid motion (either $\vec{V}$ or $-\vec{U}$) before it is deflected by the surface.

Ideally, lift and drag on any object could be calculated. One only needs to know the shape of the object, the fluid flow vector $-\vec{U}$ or $\vec{V}$, the fluid density ρ and its kinematic viscosity ζ. For each point on the object, the pressure and viscous forces are added to give the total force. This is usually an impossible task because the Navier–Stokes equation cannot be solved. However, the situation is not hopeless. Carefully chosen approximations give reasonable results.

6.2.2 Viscosity Defined

Newton was the first to quantitatively describe viscosity. Fluids characterized by Newton's viscosity equation are called "Newtonian." Air and water are Newtonian fluids. Ketchup and blood are not.

In principle, one can measure a fluid's viscosity using two flat plates of area A that are separated by a small distance, d. A fluid is placed between them and one plate slides slowly past the other with a speed, U. The "no-slip" condition of the Navier–Stokes equation means fluid in contact with either plate moves with the plate. The fluid between the two plates is sheared, so the horizontal speed depends on the vertical position, as shown in Figure 6.2.

Fluid viscosity resists the shearing with a force $F(viscous)$. For Newtonian fluids, this force is proportional to the product of the top plate speed U and the area A of the plates. The force is also inversely

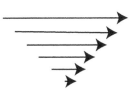

Figure 6.2 The arrows illustrate the velocity distribution of a sheared fluid. This pattern resembles the flow in a laminar boundary layer.

proportional to the separation between the plates, d. The proportionality constant that determines the magnitude of this viscous force is the same dynamic viscosity $[\rho\zeta]$ that appears in the Navier–Stokes Equation 6.2. Thus,

$$F(viscous) = [\rho\zeta]\frac{U \cdot A}{d} \tag{6.5}$$

Honey is much more viscous than water, and sailboats would hardly move in a lake of molasses. Air and water are the fluids of interest to sailors. At room temperature, their dynamic viscosities are

$$[\rho\zeta](water) \cong \frac{1}{1,000}\frac{\text{Ns}}{\text{m}^2} \tag{6.6}$$

and

$$[\rho\zeta](air) = \frac{3}{100}[\rho\zeta](water) \tag{6.7}$$

As an example, if one square meter of smooth glass were placed on a wet smooth table so that the glass and table were separated by only 1 mm of water, the glass could be moved at 1/10 m/s with a force of only 1/10 N. This is roughly the force needed to lift 10 g (about two nickels). Since water's viscosity appears to be small, perhaps it can be ignored. This would be a mistake. Even when it is small, viscosity plays a key role in determining fluid flow.

It is no surprise that air has a dynamic viscosity that is about 30 times smaller than the dynamic viscosity of water. It is hard to notice the forces on surfaces moving past one another when they are separated by only a thin layer of air.

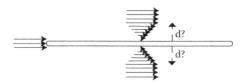

Figure 6.3   Water flows past a thin centerboard. The viscous force on the sides of the centerboard depends on the thickness d of the boundary layer.

6.2.2.1 The Centerboard Problem First Attempt   One would think that the next logical step would be to apply the definition of the viscous force to a situation important for sailors. However, as this example shows, it is more difficult to assess the viscous force for real systems than Equation 6.5 suggests.

Assume a boat is sailing directly downwind, so the fluid flow is parallel to a thin centerboard's surface, as sketched in Figure 6.3. For downwind sailing, a very thin centerboard has an insignificant pressure drag. The area on the front edge of the centerboard is so small that pressure has almost nothing to push against. But there is still a force on the centerboard because viscosity causes the centerboard to "feel" the tangential flow of the nearby water. Water touching the sides of the centerboard sticks to the centerboard, and this drags additional water with it. Small boat sailors know viscous drag is significant. Unless they are just out to enjoy the day, smart sailors raise their centerboards whenever they sail with the wind.

The obvious starting point for a calculation of the centerboard's drag is the definition of the viscous force, $F(viscous) = [\rho\zeta]UA/d$. The problem with applying Equation 6.5 is the d in the denominator. Without the other plane, what should one use for the separation d? The short answer is, "d is the thickness of the boundary layer." Outside the boundary layer, the water is nearly oblivious to the centerboard's motion. Because there is no obvious way to determine the boundary layer thickness, the calculation of $F(viscous)$ must be temporarily abandoned. This basic centerboard problem will be considered three more times. The second attempt uses Reynolds number scaling, the third attempt is based on laminar boundary layer approximations, and the final attempt considers the effect of boundary layer turbulence. All the attempts fail to give an unambiguous result for this seemingly simple viscous force problem. Fluids are complicated.

6.2.3 Viscosity Physics

Viscosity can transmit force from one plate to another, as described by Equation 6.5. It does this by transferring the motion to molecules near the plates that in turn pass them on from molecule to molecule. In water, the molecules that push on each other are near neighbors. If the water molecules can pass by each other while exhibiting a relatively small force, the viscosity is small. When the water is cold, the molecules tend to be stuck in their positions, and this allows them to push harder on each other. Thus, the viscosity of relatively sticky cold water at 10°C is 60% larger than the viscosity of warm water at 30°C. Since hot water is more "watery" than cold water, one could argue that a sailboat should be faster when sailing in warmer water. However, as will become apparent, forces are not simply proportional to the viscosity, so the difference between sailing in hot and cold water may be hard to notice.

Air is not just a thin version of water. Air's dynamic viscosity $[\rho\zeta]$ (*air*) is more mysterious because it does not change, even when most of it is removed from a container. James Clerk Maxwell was the first person to explain why "less air" could be just as viscous as "more air."

Maxwell unified electricity and magnetism. He also unified theory and experiment in his 1860s study of gas viscosity. His theoretical results on viscosity were surprising, so he performed an experiment to validate his ideas. Disks were suspended from a thin wire so they could easily rotate. The illustration in Figure 6.4 is a simplification of Maxwell's experiment. The rotating disks were placed adjacent to fixed disks so the viscosity of the air separating the disks could damp the rotation. The system was placed in an airtight container and nearly all the air was pumped out. Removing most of the air did not change the damping rate. The logical, if curious, implication of this experiment is that the dynamic viscosity of air does not depend on its density, which is exactly what Maxwell had predicted.

Maxwell's explanation looked at the microscopic origin of viscosity. The force is transmitted between the two disks by air molecules. The transmitted force is proportional to the molecular density and the ease with which the molecules can carry their motion from disk to disk. The molecules can move further when the air density is reduced because the chance of an intermolecular collision is smaller. For a

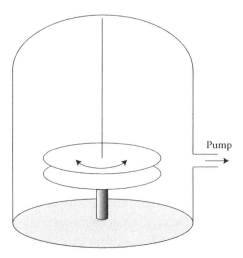

Figure 6.4   The rotating oscillations of the upper disk supported by a thin wire are damped by the viscous coupling to the fixed disk. The damping persists even when most of the air is removed.

dilute gas like air, the longer path length exactly cancels the higher density, resulting in a density-independent viscosity.

Maxwell's viscosity theory gives some insight into viscous forces when flows are turbulent and the fluid swirls about in complex paths. In a viscosity experiment, this random motion rapidly carries fluid from one plate to the other. The more effective transmission of motion from plate to plate is equivalent to a much larger viscosity. Thus, turbulent flow can be crudely characterized as a fluid with a greatly enhanced viscosity. The enhanced viscosity of turbulence generally (but not always) means more drag and less speed, so sailors often wish to minimize turbulence.

6.2.4 Viscosity, Energy, and Dissipation

Friction warms your hands when you rub them together. Viscosity does essentially the same thing, transferring a fluid's kinetic energy into heat. Just as an object sliding on a floor slows down at a rate proportional to the friction coefficient, fluid passing a fixed surface comes to a halt at a rate proportional to the viscosity. This can be seen in an experiment as simple as stirring coffee. If it weren't for viscosity, stirred coffee would keep moving forever. In practice the coffee slows by roughly a factor of three in a "decay time," τ.

The kinematic viscosity is the fundamental quantity that characterizes fluid motion, so it determines the slowing rate and the decay time. The kinematic viscosities of water and air near room temperature are

$$\zeta(water) \cong 10^{-6} \frac{m^2}{s} \tag{6.8}$$

and

$$\zeta(air) = 15\,\zeta(water) \tag{6.9}$$

Air's surprisingly large kinematic viscosity is a consequence of its low density, which is 800 times smaller than water's density. Air's density decreases with temperature, so its kinematic viscosity increases with temperature.

An estimate of the coffee cup "decay time" τ can be obtained by noting that the units of the kinematic viscosity are $\zeta = m^2/s$. Since the decay time should decrease as the viscosity increases, a reasonable guess for the decay time is

$$\tau \approx K \frac{(distance)^2}{\zeta(water)} \tag{6.10}$$

Because a squared distance has been placed in the numerator of Equation 6.10, a comparison of units shows that K is dimensionless. Using the radius of the coffee ($\cong 3$ cm) as the only available distance yields a decay time $\tau \approx K \cdot 1,000$ s for the coffee stirring. Complex swirls in stirred coffee die out quickly. But if you watch carefully, a very slow circular motion can persist for about 15 s. This suggests $K \approx [1/100 \rightarrow 1/10]$. (It is easier to see the slow decay by shaking a little black pepper into a swirling glass of water.)

Because the kinematic viscosity of air is 15 times that of water, the air circulation of an empty coffee cup should die out 15 times faster. To do this silly experiment, you need to put a top on the cup full of air after it is stirred, the deceleration is visible only if there is some dust suspended in the air.

Viscosity also damps surface water waves, which are described in Chapter 8. An accurate calculation gives a similar expression for the wave lifetime

$$\tau = \left(\frac{1}{8\pi^2} \right) \frac{\lambda^2}{\zeta(water)} \tag{6.11}$$

Here, λ is the wavelength of the water wave. The coefficient $1/(8\pi^2)$ $\cong 1/79$ is consistent with the rough estimate that the constant K is considerably less than unity.

A more complicated example of viscous damping applies to the turbulent motion of the wind, described in Chapter 9. The sun's uneven heating and the earth's rotation are constantly supplying kinetic energy to our atmosphere's circulation. Eventually, most of this energy must be turned to heat through the mechanism of viscosity. The shortest times, τ, and resulting quickest dissipation are associated with the smallest distance. This means the conversion of kinetic energy to heat occurs through the shortest-range fluctuations of the wind speed. These shortest distances are on the order of millimeters, and the corresponding shortest times are on the order of milliseconds. Although we cannot notice it, wind variations extend to a nearly microscopic scale.

6.3 Reynolds Number

At this point, one should be convinced that even rough estimates of fluid forces are difficult to obtain. Any trick that could simplify these difficult problems would surely be appreciated. The Reynolds number is the most important labor-saving trick of fluid mechanics.

Suppose one wanted to know the fluid drag on a variety of spheres—big ones and little ones. Assume one was also curious about the drag on these spheres for a range of fluid speeds, U. The drag phenomenology from Chapter 2, produced the formula

$$F_D = \frac{C_D}{2}\rho \cdot A \cdot U^2 \tag{2.6}$$

This appears to give the answer, provided drag coefficient C_D is known. Unfortunately, fluids are not simple. The drag coefficient is not really a constant. It can change with the fluid speed (U or V), the area A, and the fluid density ρ. But a sphere is a sphere, so one expects some simplification. That is exactly right, and the Reynolds number tells one how to save time and money.

There are (at least) two ways to derive the Reynolds number: the careful method and the tricky method. The careful method shows that the Navier–Stokes equation is unchanged when various quantities are properly scaled. The careful method takes longer, so the tricky method is presented here.

6.3.1 Reynolds Number Defined

The trick used to derive the Reynolds number is based on units and the formula for the drag. It is a generalization of the well-known rule that forbids a comparison of apples and oranges. Assume a fluid with density ρ moves past a sphere of diameter L at a speed U. The cross-sectional area of the sphere is $A = \rho(L/2)^2$. Using this formula for the area, the formula for drag from Equation 2.6 becomes

$$F_D = \frac{1}{2}C_D\rho\pi\left(\frac{L}{2}\right)^2 U^2 \qquad (6.12)$$

Consider the units in Equation 6.12.

1. The drag force F_D: kilograms times meters divided by seconds squared.
2. The fluid density ρ: kilograms per cubic meter.
3. The fluid speed U: meters per second.
4. The sphere's diameter L: meters.
5. The kinematic viscosity ζ: meters squared per second.

A comparison of the units shows that the drag coefficient C_D in Equation 6.12 must have no units at all. It is a "dimensionless" number. The drag coefficient can depend on all the physical quantities, but they must be combined in such a way as to be dimensionless. The *only* dimensionless quantity one can construct from ρ, U, L, ζ is the Reynolds number

$$R = \frac{U \cdot L}{\zeta} \qquad (6.13)$$

Even though the drag coefficient may not be constant, it can depend only on the Reynolds number:

$$C_D \rightarrow C_D(R) \qquad (6.14)$$

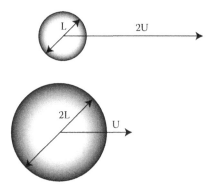

Figure 6.5   The big sphere of diameter L and speed U has the same Reynolds number (and the same drag) as the smaller sphere with diameter L and speed $2U$.

Now, one can compare the drag on the big and little spheres shown in Figure 6.5. The sphere with twice the diameter is moving at half the speed, so the two spheres have the same Reynolds number. They also have the same drag because the doubled velocity increases the drag by a factor of four, but half the diameter decreases the area by a compensating factor of four.

The Reynolds number is not restricted to spheres, and it is not restricted to drag. It characterizes flow around any object. For example, one can use the Reynolds number to compare the lift and drag on a variety of centerboards (big ones and little ones) over a wide range of fluid speeds. The Reynolds number also allows a comparison of the lift and drag on objects in two different fluids—air and water, for example.

For objects other than a sphere, L must be changed from the sphere diameter to more general length that characterizes the size of the object. Without spherical symmetry, C_D and C_L also depend on the direction of fluid flow. However, no matter how complicated the shape of an object, its drag and lift coefficients can only depend on the Reynolds number.

The Reynolds number invariance means scale models can be used to determine C_L and C_D on systems where direct measurement would be difficult. Results obtained from a half-sized model of a sail or a centerboard can be trusted, provided the fluid speed is doubled so R is unchanged.

It is unfortunate that the Reynolds number scaling cannot be applied to boats that float on the water. Doubling the speed while halving the length keeps the Reynolds number the same, but the wake produced by a 10-m boat traveling at 5 m/s is very different from the wake produced by a 5-m boat traveling at 10 m/s.

6.3.1.1 The Centerboard Problem: Second Attempt   Return to the problem of the viscous force on the thin centerboard. Assume the centerboard extends to a depth, L, in the water and assume its width is $L/2$. That means its surface area (counting both sides) is $A = L^2$. The drag can then be written as

$$F(viscous) = \frac{1}{2}C_D(R)\rho(water)L^2 \cdot \qquad (6.15)$$

This isn't really an answer because the drag coefficient C_D is an unknown function of the Reynolds number $R = U \cdot L/\zeta$.

The following guess is another example of plausible but incorrect physics. It highlights the deceptive nature of fluid mechanics.

Guess: It seems reasonable to assume the viscous force should be proportional to the viscosity. The linear dependence of $F(viscous)$ on viscosity can only be achieved by making the drag coefficient proportional to the viscosity. This is accomplished only if $C_D = 2K/R$, where K is a constant and R is the Reynolds number. Using this guess gives

$$F(viscous : wrong) \propto \zeta \cdot L \cdot U \qquad (6.16)$$

The Reynolds number scaling means one can obtain a viscous force that is proportional to the viscosity only if one abandons the requirement that the force is proportional to the surface area and the square of the velocity. For sailboats, the forces are generally proportional to surface areas and squared velocities. So this is another failed attempt to solve the centerboard problem.

The centerboard problem is just one example of the peculiar nature of fluid forces. The conditions apply to drag or lift, pressure of viscous forces, hulls, or sails. For example, because the downwind force on a sail is $F = C(R)AV^2/2$, any one of the following three statements implies the other two.

1. The force on a sail is proportional to the sail area, A.
2. The force on a sail is proportional to the square of the apparent wind speed, V.
3. The force on a sail is independent of the viscosity.

These statements are all equivalent because they all imply a constant drag coefficient. One generally expects the lift and drag forces on an object to be proportional to the surface area of that object. The condition that $F \propto A$ means the force must be proportional to the square of the velocity and independent of the viscosity.

This appears to be a real dilemma for the viscous force on the centerboard. How can there be a viscous force that is independent of the viscosity? What would happen to this force when the viscosity vanishes? Fluids are tricky.

6.4 Boundary Layers

A fluid "feels" the viscous force from a surface primarily within the boundary layer. Outside this layer, the fluid is nearly oblivious to the viscous surface force.

Boundary layers come in two forms: "complicated" and "very complicated." The complicated boundary layers are laminar and are attached to the surface (sail, centerboard, rudder, hull, etc.). They become very complicated when turbulence develops and/or the layer separates from the surface. "Laminar" means all regions in the layer flow in about the same direction. If separation does not occur, this flow is parallel to the surface. "Turbulent" means the velocity within the boundary layer has a large random component. In a turbulent boundary layer, fluid can flow toward and away from the surface. To add complexity, separation and turbulence are intertwined.

An oversimplified discussion of the laminar boundary layer will be followed by even more flimsy discussions of turbulence and separation.

6.4.1 Laminar Boundary Layer

A simple laminar boundary layer is a transition region. Fluid sticks to the surface on the inside of the boundary layer and changes smoothly to nearly viscous-free motion at the outside. The fluid flow in the layer is similar to the shear flow shown in Figures 6.2 and 6.3, with flow vanishing at the surface. The flow speed is roughly U or V at its outer edge.

The laminar boundary layer is initially very thin, but it thickens as the fluid moves along a surface. If d is the boundary layer thickness, and t is the time the fluid has traveled along the surface, one can roughly estimate the time-dependent boundary layer thickness $d(t)$.

For simplicity, consider a flat surface parallel to the fluid flow $\vec{U}$ like the centerboard in Figure 6.3. The fluid near the surface can be treated as a set of ultra-thin sublayers parallel to the surface, each with a tiny thickness δ. Only the first sublayer touches the surface, so initially (small t) viscous forces affect only the motion of this first sublayer. As the first sublayer slows, it drags the second sublayer, slowing its motion as well. This process repeats to sublayers 3, 4, ... , n. The boundary layer consists of all n sublayers, so its thickness is $d = n\delta$. The fluid speed changes smoothly from zero at the surface to nearly U at the outermost sublayer n. That means the speed differences between adjacent sublayers is approximately U/n. The surface interacts only with the innermost sublayer, so the force on the surface is proportional to the speed of the first sublayer, which is also roughly U/n. Since $d = n\delta$, a proportionality is

$$Friction \ \ force \propto \frac{U}{d} \qquad (6.17)$$

The rate at which the boundary layer grows is proportional to the friction force on the surface, so

$$\frac{\text{Change in } d}{\text{Change in } t} \approx \text{Growth Rate} \propto \frac{U}{d} \qquad (6.18)$$

The square root function $d(t) \approx K\sqrt{t}$ is a solution to this equation. As shown in Figure 6.6, it grows rapidly when d and t are small. Later, it grows more slowly.

A quantitative estimate of $d(t)$ is obtained by examining the units of the constant K. After passing by a surface for a time t, the only dimensionally correct expression involving the physically relevant V, ρ, t, and ζ that scales with the square root of the time is

$$d \approx \sqrt{t\zeta} \qquad (6.19)$$

This means a diffusion-like mechanism determines the boundary layer growth, with the kinematic viscosity ζ playing the role of the diffusion constant.

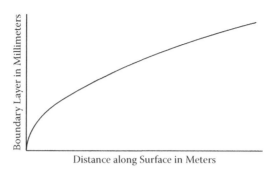

Figure 6.6 The boundary layer thickness d increases as the square root of the distance along a surface. The different scales (millimeters versus meters) emphasize the small size of the boundary layer.

An example illustrates the filmy thinness of a typical laminar boundary layer. The kinematic viscosity of water is $\zeta = 10^{-6}\,\mathrm{m^2/s}$. For a centerboard with a horizontal span $L = 0.5$ m and the boat speed $V = 5$ m/s, the laminar boundary laminar layer at the back edge of the centerboard should be about 1/3 mm. The boundary layer is even thinner near the front. The same Equation 6.19 for $d(t)$ applies to the growth of the boundary layer on a sail. Air's kinematic viscosity is 15 times that of water and a typical sail span is 5 m, instead of half a meter. Thus, the laminar boundary layer would be closer to 3 mm at the back edge of a sail. However, as is shown in Chapter 9, the wind is turbulent, so the relevance of a laminar boundary layer calculation for sails is questionable. Actually, there are many sailing conditions where the water is also turbulent, so all these results should be approached with caution.

6.4.1.1 The Centerboard Problem: Third Attempt An approximate expression for the thickness of a laminar boundary layer allows another attempt to find the viscous force on the centerboard. The first attempt was aborted because the viscous force depends on the unknown interplane distance, d. Now that we have an estimate, it is reasonable to approximate d by the boundary layer thickness.

Combining $F(viscous) = [\rho\zeta]U \cdot A/d$ from Equation 6.5, $d \cong \sqrt{\zeta t}$ from Equation 6.19 and $t \cong L/U$ [distance is speed times time] gives

$$F(viscous) \cong 2\rho A U^{3/2}\sqrt{\frac{\zeta}{L}} \qquad (6.20)$$

The factor 2 appears because the boundary layer is thinner toward the front of the centerboard.

This result is consistent with the requirement that the drag coefficient depends only on the Reynolds number. For this result,

$$C_D = \frac{4}{\sqrt{R}} = 4\sqrt{\frac{\zeta}{LU}}$$

In this attempt to solve the centerboard problem, the force is proportional to the square root of the kinematic viscosity instead of being just proportional to the viscosity (as was the case in the second attempt). The square root appears because a larger viscosity leads to a thicker boundary layer, and these effects partially cancel.

This third attempt is still not satisfactory. For water, $\rho = 1,000$ k/m^3 and $\zeta = 10^{-6}$ m^2/s. Using these values,

$$F(viscous) = K\frac{AU^{3/2}}{\sqrt{L}} \tag{6.21}$$

The constant is $K \cong 2$ N-s$^{3/2}$/m^3. For a boat similar to a Thistle, the area of a centerboard (both sides) is a little less than a square meter and the horizontal length of the centerboard is also less than a meter. Using these values, the force (in Newtons) is roughly twice the three-halves power of the boat speed (in meters per second). For the Fresh Breeze, $U \cong 5$ m/s, and this gives a force of roughly 20 N. This estimate is probably too small, but it is in the right ballpark.

This third attempt to find the force on the centerboard presents other problems. The Reynolds number scaling means the force is not proportional to the square of the boat speed, and it is not proportional to the surface area of the centerboard. Also, because the force is proportional to $A/\sqrt{L}$, it appears that a shallow centerboard with a large horizontal span would produce less drag than a deeper centerboard. In practice, the centerboard on the left of Figure 6.7 would probably have less drag.

There is still some missing physics. The problem lies in the instability of laminar boundary layers. They generally turn into turbulent boundary layers. Turbulence in boundary layers increases the viscous force because fluid motion can be rapidly transmitted to and away from surfaces.

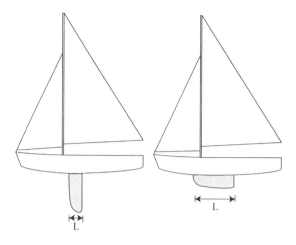

Figure 6.7   Two similar sailboats with differently shaped centerboards. A laminar boundary layer would produce less drag on the centerboard on the right. The centerboard on the left is actually preferred.

6.4.2 Turbulence Basics

Turbulence and separation put an end to the life of the simple laminar boundary layer. The transition to turbulence is very important for sailing, so we should really understand it. But there is a problem. The word "turbulent" is almost a synonym for "incomprehensible."

Turbulence is characterized by unpredictable random and disorderly flow. It is everywhere. It can be seen when a rising column of smoke changes from orderly to complex. It can be seen when the stream from a faucet breaks into disorderly droplets. Our atmosphere is turbulent, and this turbulence helps provide reasonable temperatures and relatively clean air. Without turbulence, life as we know it would be impossible.

The tiny size of the boundary layers surrounding a sailboat makes their observation difficult. If one is willing to ignore a lot of details, our atmosphere is a giant-sized turbulent boundary layer. We live in this atmospheric boundary layer, so it should be much easier to observe. Chapter 9 presents some measurements and a theory of the turbulent wind near the earth's surface. All sailors know that wind is chaotic and unpredictable, so it should be no surprise the fluid motion in the thin turbulent boundary layers rubbing on sailboats is equally confusing.

A good theory of turbulence should predict its onset. Since no good theory is at hand, we rely on the Reynolds number as a rough criterion.

A "big" R on the order of a million or more implies turbulence. A "small" R less than 100 means fluid flow is dominated by viscosity. Between big and small R lies a transition region. In this midrange, fluid behavior can be quite variable and surface smoothness can make a significant difference. An example is the comparison in Section 7.3 of the drag on a smooth sphere and a golf ball.

Although there are exceptions, one generally expects turbulence to appear at a threshold Reynolds number $R \approx 10^5 \rightarrow 5 \times 10^6$. This wide range for R is not carelessness. Streamlined objects need a much larger R to develop turbulence. Combining the definition $R = U \cdot L / \zeta$ with $\zeta(water) = 10^{-6}\, m^2/s$ and the turbulence threshold gives an estimate of the length L required for turbulence.

$$L(water; \text{ in meters}) \approx \left(\frac{1}{10} \rightarrow 5 \right) \frac{1}{U(\text{in meters/second})} \qquad (6.22)$$

Because the kinematic viscosity of air is 15 times that of water,

$$L(air; \text{ in meters}) \approx \left(\frac{3}{2} \rightarrow 75 \right) \frac{1}{V\left(\text{in meters/second}\right)} \qquad (6.23)$$

For a Fresh Breeze where $U \cong V \cong 5$ m/s, this estimate suggests that boat hulls should excite turbulent motion. When winds are very light, a small U can make $L(water)$ comparable to or even larger than a boat length. Then turbulence is not a problem. This makes sense. When a sailboat is barely moving, one does not envision a significant turbulent contribution to drag. The estimate of $L(air)$ means turbulence should be less likely for the wind on sails. Since wind is generally turbulent before it hits the sails, this estimate does not carry much significance.

One can also estimate Reynolds numbers for different parts of a sailboat. The Reynolds numbers for a centerboard, rudder, or jib (with smaller values for L) are smaller than the corresponding numbers for the entire boat. If one calculates the Reynolds number for the front third of a centerboard, one may (properly) conclude that turbulence develops only after the water has flowed past a surface for some distance.

A famous example of complex behavior in the transition range is the significant decrease in the drag coefficient of a sphere for a Reynolds number $R \approx 10^5$. This change in C_D, which is called the "drag crisis,"

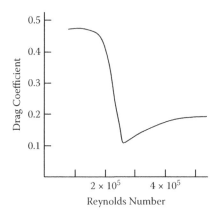

Figure 6.8   The variation of the sphere's drag coefficient for a range of Reynolds numbers shows a drag crisis. An object with a different shape can have a much different drag, and it may not exhibit a drag crisis.

is caused by a fairly abrupt change in the fluid flow pattern when turbulence starts to dominate the fluid motion.

The drag crisis is far from universal because the shape of an object has a significant influence on the fluid flow. There is no drag crisis for downwind sailing because the wind behind the sail is always turbulent. The drag coefficient is very sensitive to even a small amount of initial turbulence, so it is difficult to measure accurately. The curve in Figure 6.8 is an average of different measurements.

6.4.3 Turbulent Boundary Layer

Because turbulence requires a fairly large Reynolds number, boundary layers are usually born laminar and grow up to be turbulent. As the fluid moves a distance L along the surface, the "local" Reynolds number increases to a value where turbulence cannot be avoided. When the laminar motion becomes turbulent, fluid near the surface swirls about, moving back and forth within the boundary layer. Fluid from the outer edge of the boundary layer with speed U can make frequent visits to the surface. The increased communication between the surface and the outer regions of the boundary layer means the surface is subjected to larger friction forces. It can be approximated by an effective turbulent viscosity that is much larger than the real kinematic viscosity ζ, as mentioned in Section 6.2.3.

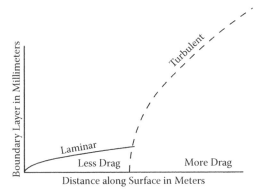

Figure 6.9   The turbulent boundary develops as fluid moves along a surface. The boundary layer thickness and the viscous force are increased by turbulence.

For a laminar boundary layer, Equations 6.19 and 6.20 show that boundary layer thickness and surface drag are both proportional to the square root of the kinematic viscosity ζ. If turbulence can be characterized by a greatly enhanced ζ, then perhaps the laminar results can be adopted with just one change; a much larger viscosity. This simple magnification of viscosity when turbulence occurs is sketched in Figure 6.9. It is observed that both the boundary layer thickness and the viscous force increase by up to an order of magnitude when the boundary layer becomes turbulent. Streamlined designs and smooth surfaces that retard the development of turbulence are clearly desirable. Once turbulence is initiated, it grows rapidly. Sometimes this means that a single rough point on an otherwise smooth surface could start turbulence "before its time."

The approximation that says a turbulent boundary layer is just a laminar boundary layer with a greatly enhanced viscosity is a gross oversimplification. The velocity profile of the turbulent boundary layer is different from that of the laminar layer. Fluid speeds in a laminar boundary layer increase roughly linearly with distance from the surface, as suggested by Figure 6.2. The velocity increase in a turbulent boundary layer probably looks more like the idealized altitude dependence of wind speeds in our turbulent atmosphere. This logarithmic dependence is discussed in Section 9.3 and is sketched in Figure 9.6.

6.4.4 Boundary Layer Separation

"Separation" really means "early separation." When a fluid flows past a surface, the boundary layer must eventually leave the surface. Early separation means the boundary layer departs before it gets to the back edge. When a fluid separates from a centerboard or sail before it gets to the stern end, fluid is sucked in from both sides of the separation point. That means surface fluid that is behind a separation point will be pulled forward. Gross examples of separation are readily observed. People who ride in convertibles may notice that the wind is blowing their hair forward (if they still have hair). This seems like the wrong direction. You may notice that motorcyclist's shirts are blow up and forward. This may seem wrong, too (and sometimes in bad taste), but you can't argue with observation. The passenger's head in the convertible and the motorcyclist's back are both behind the separation point.

Separation can also be observed on sails when sailing upwind. If a piece of light material is attached to the leech (back edge) of a sail, it indicates an average local wind. When this wind indicator is blown forward to the sail's leeward side, it means the separation point on the sail has moved to a point forward, as is sketched in Figure 6.10. Sailors often feel that this separation should not occur. Even though this seems to be a sensible rule for upwind sailing, it is frequently violated with apparently little penalty.

Finally, there is the interaction between turbulence and separation. Ordinarily, turbulence and separation both increase drag. But turbulence increases the thickness of the boundary layer and thicker boundary layers are more firmly attached to a surface. I have no simple explanation of the "sticky" turbulent boundary layer, but it is

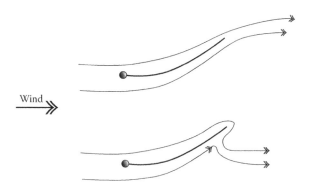

Figure 6.10   Wind flow at the top leaves the back edge of the sail. At the bottom separation occurs on the lee side. The wind paths shown are a sketch and are not the result of any calculation or measurement.

confirmed by experiments and numerical simulations. Even though the mechanism is difficult to understand, turbulence can inhibit separation and decrease drag. This is the explanation of the "drag crisis" shown in Figure 6.8.

6.4.4.1 The Centerboard Problem: Final Attempt The turbulence and separation that characterize most real boundary layers make a quantitative calculation of the centerboard drag nearly impossible. The final conclusion is far from a solid result. The drag is probably significantly larger than the 20-N estimate obtained using a laminar boundary layer. The effective viscosity of turbulence is not necessarily proportional to the viscosity of the fluid, so it is hard to say how the drag varies with the viscosity of the water, the speed of the boat, or the surface area of the centerboard.

In short, after all this work, I have essentially given up the quest for a simple physical answer for the viscous drag on the centerboard. Experts in the techniques of numerical fluid dynamics can obtain reasonable results.

6.4.4.2 Problems Harder than the Centerboard Problem If one were capable of finding both pressure and viscous forces, one could extend a drag calculation to a keel instead of a centerboard. A keel, typically shaped like an elongated egg, is subject to both pressure drag and friction drag. The pressure on the front of the keel is larger than pressure on the back. Also, water moving past the edges of the keel exerts a viscous drag. Summing the two gives the total drag.

6.5 Euler Equation

The Euler equation ignores viscosity. This nonviscous idealization of the Navier–Stokes equation allows one to "prove" some important relations. These include fundamental concepts named after d'Alembert, (Daniel) Bernoulli, and Kutta–Joukowski (the latter is two people). These are both useful and confusing tools for explaining lift and drag.

The utility of the Euler equation is surprising, considering the essential role of viscosity. A key to applied fluid mechanics is a selective use of the Euler equation. Although it fails near surfaces where viscosity is important, it is very useful far from surfaces and wakes (e.g., wind shadows). A clever merging of the Euler equation with the Navier–Stokes equation can yield accurate results.

Removing the troublesome viscosity term from the Navier–Stokes equation gives the Euler equation.

$$\frac{D\vec{u}}{Dt} = -\frac{1}{\rho}\vec{\nabla}p \qquad (6.24)$$

Here, as previously, $\vec{u}$ is the fluid velocity and $D\vec{u}/dt$ is the acceleration. The fluid density is ρ, and p is its pressure. The local velocity $\vec{u}$ near a surface can be quite different from the distant velocity $\vec{U}$ or $\vec{V}$, which is either the velocity of the boat through the water or the velocity of the wind with respect to the sail.

Since viscosity is absent, the Euler equation also abandons the no-slip boundary condition. Surfaces are no longer sticky, and the idealized nonviscous fluid can slide unimpeded along a surface. In the absence of viscosity, the forces on a surface (sail or hull) are supplied only by the pressure, and this force is always perpendicular to the surface.

Concepts that follow from the Euler equation are briefly described here. The results clearly show that viscosity cannot be completely ignored.

6.5.1 d'Alembert's Paradox

This paradox is disconcerting because it says there should be no lift and no drag. If true, it would also mean no sailing. A simple version of the paradox follows from the symmetry of the Euler equation for the restricted steady-state case where the velocity does not vary in time ($\partial\vec{u}/\partial t = 0$). Using Equation 6.3, $D\vec{u}/Dt \rightarrow (\vec{u}\cdot\vec{\nabla})\vec{u}$ in the steady state. Since the acceleration term is now proportional to the squared velocity, a reversal of $\vec{u}(\vec{r})$, so $\vec{u}(\vec{r}) \rightarrow -\vec{u}(\vec{r})$ changes nothing. This means the pressure is also unchanged when the velocity is reversed. The same pressure means the same drag force. Physically, we know drag must be reversed when $\vec{V}$ is reversed. These contradictory statements can be resolved if the drag is zero because the only vector that is the same when it is reversed is the zero vector. A generalization of this paradox says that both lift and drag should vanish. That's the essence of the paradox, but there is an unusual complication for fluid motion confined to two dimensions. Despite complications, d'Alembert's paradox spectacularly demonstrates the danger of ignoring viscosity.

6.5.2 Bernoulli's Equation

The famous Bernoulli's equation

$$\frac{1}{2}\rho u^2 + p = \text{constant} \qquad (6.25)$$

says the kinetic energy density plus the pressure of a fluid is a constant. This equation can only be derived from the Euler equation, where viscosity is ignored. External forces like gravity are also ignored in Equation 6.25. As before, u is the fluid speed, ρ is its density, and p is the pressure.

One can roughly justify Bernoulli's equation through the example of a fluid flowing through a tube, as shown in Figure 6.11. If the tube has a constricted section, the fluid must speed up to make it through the constriction. An analogy is road repair. As cars approach a construction zone where the freeway narrows from two lanes to one lane, traffic crawls along. But once cars have entered the single lane, drivers stomp on the gas and accelerate to twice their former speeds. A fluid entering the tube constriction experiences a similar acceleration. A pressure difference supplies the force that drives the fluid ahead. The pressure is used up, resulting in a lower pressure in the region where the speed is larger.

Bernoulli's equation applies for either direction of fluid flow, so the pressure change should be the same for fluid flowing into or out of a constricted region. An attempt to test Bernoulli's equation by blowing through a constricted tube mounted with pressure gauges often yields disappointing results. The pressure variations are only approximately described by Bernoulli's equation because viscosity has been ignored.

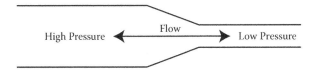

High Pressure ←——— Flow ———→ Low Pressure

Figure 6.11   Bernoulli's equation says pressure should be larger when the fluid velocity is smaller, and the result applies for either direction of fluid flow.

In many situations, Bernoulli's equation is a reasonable approximation. It is frequently used to explain lift, most notably on airplane wings. Carburetors, aspirators, birds, bats, and curve balls are explained in part by Bernoulli's equation. In all these examples, one associates reduced pressure with faster fluid motion.

The logic behind lift is confusing. It appears that Bernoulli's equation can explain lift. The argument is that the air on the top of a wing or the leeward side of a sail is moving faster than the air on the bottom of the wing or the windward face of the sail. But Bernoulli's equation and d'Alembert's no-drag and no-lift paradox are both derived from the Euler equation. How can the Euler equation deny lift and also produce an equation for its calculation? The Kutta–Joukowski theorem relating circulation and lift does little to resolve the paradox.

6.5.3 Circulation

Cyclones and whirlpools are intuitive examples of circulation, but fluids with circulation do not always travel in circles. Circulation has a formal definition in terms of an imaginary insect. This insect quickly flies a closed path through the air (or water). Along the path, this bug may experience headwinds and tailwinds, and this will influence the time needed to make the trip. The total net tailwind the bug experiences when executing a counterclockwise closed path is called the "circulation" Γ of the path. Formally, it is defined as

$$\Gamma(path) = \int_{path} \vec{u} \cdot d\vec{r} \qquad (6.26)$$

Fluid motion is "irrotational" if the circulation around every path is zero. The shear motion shown in Figure 6.12 is rotational because a bug's counterclockwise flight path would meet more wind resistance than a clockwise path.

Fluid circulation is caused by a twisting or shearing action. In fluids, this shear can only be produced by viscosity. Since viscosity is eliminated from Euler's equation, one can prove that circulation should not change in time. In particular, Euler's equation says irrotational motion should remain irrotational forever.

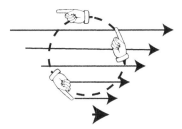

Figure 6.12   In this shear flow, traveling the counterclockwise circle will take longer, so the circulation is clockwise.

6.5.4 Kutta–Joukowski Theorem

The Kutta–Joukowski theorem relates lift to circulation. The result is consistent with Bernoulli's equation. In upwind sailing, the wind moves faster on the leeward side of a sail and slower on the windward side. Bernoulli's equation then implies there will be a pressure difference between the two sides, and this difference produces the lift. Circulation also describes the lift. If the wind speed is different on the two sides of the sail, the bug flying around the sail will experience net headwind or tailwind. This means there is circulation around the sail. One can use Bernoulli's equation to express the lift in terms of the circulation

$$lift \approx \rho\, \Gamma\, (around\ sail)V \qquad (6.27)$$

Again, this makes one stop and wonder. If there is circulation there should be lift, but d'Alembert says there is no lift. Hence, the circulation around a sail should vanish if viscosity were really zero. The circulation vanishes because the Euler equation says an irrotational fluid should remain irrotational forever.

6.5.5 Lift's Many Explanations

There is a more intuitive and basic explanation of lift that is independent of any assumptions about viscosity. The sail experiences lift because the sail deflects the wind. Equal and opposite forces mean that if the sail pushes on the wind in one direction, the wind pushes the sail the other way. This explanation, which relies only on the rule of equal and opposite forces, is surely correct.

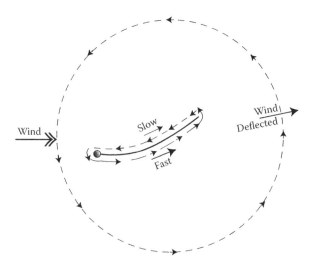

Figure 6.13 Lift occurs when a sail deflects the wind. The deflected wind implies circulation and a slower velocity on the upper side of the sail. All of these are associated with lift.

The several explanations of lift are closely related, as illustrated in Figure 6.13. If a sail deflects the wind, a large circle drawn around the sail will show circulation because the deflected $\vec{u}$ downwind is partially aligned with a large circle used to calculate circulation. The nonzero circulation means the Kutta–Joukowski theorem also describes the lift.

The circulation around the big circle and the inner loop will be identical if the fluid between them is irrotational. Euler's equation certainly suggests that the flow between the inner circle and outer loop should be irrotational. If there is circulation on the inner loop, the wind must be moving faster on the leeward (bottom) side of the sail. This is consistent with the lift predicted by Bernoulli's equation.

How is this consistent with d'Alembert's paradox? The simple answer is that the Euler equation requires the circulation around the sail to vanish. Then there is no lift, no circulation, and no sailing. One must rely on viscosity to play its role in establishing circulation, and this can only be understood through the Navier–Stokes equation. The dynamics that establish circulation are complex, involving the generation of whirlpools (vortices) and downwind flows called "wakes."

Intuition can only take one so far in understanding how viscosity produces circulation and lift. Fred, the clever sailor, covers the windward side of his sail with sandpaper. This slows the air on that side. Bernoulli's equation says the slower velocity would increase the pressure. Kutta–Joukowski says there should be more circulation. These mean more lift. Fred is not that clever. Covering one side of your sail with sandpaper won't help.

6.5.6 Two Dimensions

Two-dimensional systems are only idealizations, but they provide insight and a basis for approximations. Two-dimensional drawings of three-dimensional systems, like Figure 6.13, hide many problems because real circulation has three-dimensional structure.

The two-dimensional world provides an escape from the no-lift part of d'Alembert's paradox, but the no-drag conclusion is left intact. The lack of lift is related to vanishing circulation, but in two dimensions the core of the circulation can be stuffed inside the sail (or centerboard). Circulation around any loop that does not contain the sail vanishes, so the fluid is irrotational even though circulation around the sail means there is lift. This doesn't work in three dimensions.

A famous and soluble example of lift in two dimensions is illustrated in Figure 6.14. Both flows are solutions to the Euler equation, and both velocities $\vec{u}$ approach the constant value $\vec{U}$ at a large distance. However, the fluid flows up and over the circle on the left and does the opposite for the other circle. The opposite circulations around the circles means the lift force F_L is "up" on the left and "down" on the right. The lines around the circles represent fluid paths, but no arrows are drawn because the same lift (and vanishing drag) is obtained for fluid flowing either left to right or right to left.

The existence of two fluid flows and two different lifts for the same object and the same $\vec{U}$ is not physically satisfying. Nonunique solutions

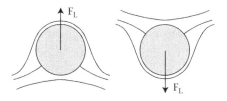

Figure 6.14   Two-dimensional flows around a circle. The lines indicate fluid paths that can be either left to right or right to left. For the Euler equation, there is no drag but lift is "up" on the left circle and "down" on the right circle.

are another perplexing aspect of the Euler equation. The ambiguity is eliminated by viscosity. In the real world, one can use viscosity to control the circulation around these circles. The sticky-surface boundary condition of the Navier–Stokes equation means a rotating circle would produce lift by making the fluid follow the motion of the surface. However, the fluid flow would exhibit only a rough similarity to the flows shown in Figure 6.14. Boats using rotating cylinders have been constructed that successfully harness the wind's power by dragging the fluid around in the right direction. Curve balls use roughly the same idea.

The possibility of obtaining lift of either sign is not limited to circles. It applies to two-dimensional sails and centerboards in the absence of viscosity. The sensible arguments that attributed lift to the sail shape in Figure 6.13 would fail if one could ignore viscosity and the sail were two dimensional. In a two-dimensional nonviscous world, there would be no drag on the sail. That is only mildly disturbing. It is more disturbing that the lift force could be in the sensible direction (down) or it could be reversed, because the Euler equation cannot distinguish the direction of the fluid flow. Our intuition tells us that the lift should be reversed if the flow is reversed, but the Euler equation doesn't notice. Viscosity again comes to the rescue. For a streamlined shape like a sail (or a centerboard), viscosity and common sense demand that the disturbed air wake (wind shadow) lie on the downwind side of a sail. With this condition, reasonable solutions to the Euler equation are obtained. A famous but complicated calculation for a perfectly flat surface in the two-dimensional world gives the lift coefficient as

$$C_L \cong \pi \sin(2\theta) \qquad (6.28)$$

Since $\sin(2\theta)$ is proportional to the angle of attack for small θ, this is a key result. It is needed to explain why sails and centerboards can produce large lift-to-drag ratios, and the result was incorporated in the phenomenology of Equation 3.10. As described in Section 3.3.3, an extension of Newton's impact theory fails to produce this result.

The two-dimensional examples suggest a way to construct sails (and centerboards) with very large lift-to-drag ratios. Just make the sails much taller than they are wide, so they are almost two dimensional. Almost two-dimensional sails could almost be described by two-dimensional fluid mechanics. Then, if viscosity is ignored, lift without drag is almost possible. Gliders and albatrosses are made with very long narrow wings for this purpose. Flat end-plates are attached to some airplane wings. These plates force the air to act two-dimensionally because flow around the ends is discouraged. For sailors, there are practical limitations to very tall sails and centerboards. Capsize is an obvious concern for a sailboat with a very tall mast.

6.6 Why Are Fluids So Complicated?

Almost every seemingly simple fluid mechanics question turns out to be impossibly difficult. Simple questions reveal ambiguities, para-doxes, and contradictions. Two perplexing but practical examples were considered here. They are the viscous force on the thin center-board and the characterization of lift.

Attempts to find a simple expression for the viscous drag on the cen-terboard were discouraging. The first attempt was abandoned because the viscosity formula only applied for thinly separated surfaces, not for a single surface. The second attempt gave a force proportional to viscosity. But the Reynolds number means a force proportional to the viscosity cannot be proportional to the centerboard area. This is not satisfactory. The third attempt, based on laminar boundary layers, suggested that centerboards should have unreasonable shapes. The final attempt admitted that there is no easy answer because turbulent boundary layers are so complicated.

Lift is even more difficult than drag. To simplify, one is moti-vated to consider the Euler equation that eliminates the viscous term from the Navier–Stokes equation. A judicious application of the Euler equation leads to important principles, like the Bernoulli equation, that can be used to evaluate lift. However, there are troublesome para-doxes that frustrate attempts to really understand lift.

Not all of fluid mechanics is discouragingly difficult or ambiguous. There is one solid result whose rigor and precision can be trusted. That is the Reynolds number scaling. However, for sailors even this result is of limited use because the Reynolds number cannot be applied to surface waves and wakes.

7

SURFACES

Racing sailors are often obsessed with producing the smoothest possible hulls, centerboards or keels, and rudders. Sails should be smooth, too. It seems sensible to assume a smoother surface would be subjected to a smaller viscous force. How smooth is a typical sailboat hull? Is this smooth enough? Is a smooth surface always the best, and if so why? These smoothness questions are trickier than one might think. Some curiosities make one think twice about surfaces.

A familiarity with small distances is needed to judge roughness. A micrometer, μm or 10^{-6} meters, is an appropriate distance to consider. Fluid mechanics suggests that 10 μm is "smooth enough," but many sailors would not be happy with this roughness. Ten micrometers is about one-seventh the thickness of a human hair or one-fifth the thickness of a standard piece of cellophane tape. This is on the borderline of visibility. Sometimes roughness can be felt even if it can't be seen. Under ideal conditions a human finger is amazingly sensitive and can detect a surface bump not much more than 1 μm high.

There is a difference between a smooth surface and a shiny surface. The wavelength of visible light is about (1/2) μm, and any structure on this small distance scale determines whether the surface is dull or shiny. Although a mirror-like shine is not important, visible bumps and scratches on sailboat surfaces are likely suspects for the instigation of turbulence and increased drag.

7.1 An Example

The hulls of many popular sailboats have an outer polyester resin coating called "gelcoat." There are many variations of gelcoat. One example, polyethylene terephthalate, is shown in the Figure 7.1. The

169

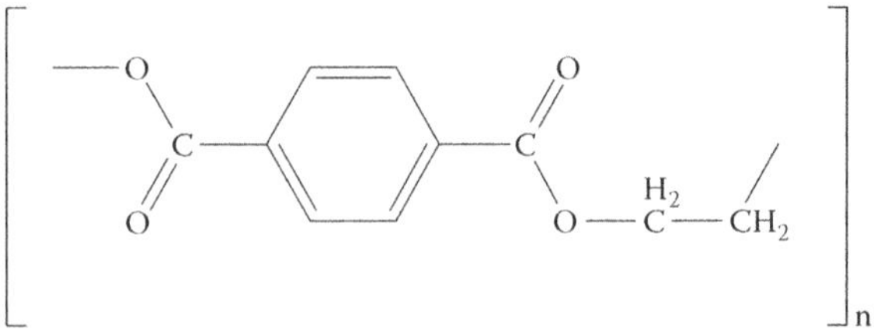

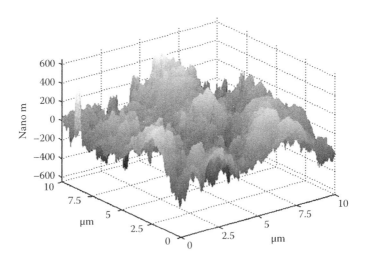

Figure 7.1 An example gelcoat polymer use to make a smooth sailboat surface.

hexagon is populated by carbon atoms. The little subscript "n" means this unit is repeated many times because it is a polymer.

Roughness measurements of two gelcoat surface pieces removed from a Flying Scot sailboat are shown in Figures 7.2 through 7.5. The perspective views look like mountain ranges because the vertical scale is 15 times magnified compared to distances on the surface. Actually, all dimensions in Figures 7.2–7.5 are tiny. The scanned areas are only 10 μm squares. The magnified vertical scales are in nanometers. One nanometer is one-thousandth of a micrometer. This means the highest peaks and valleys are about (1/2) μm high.

Figure 7.2 The two-dimensional profile of a small piece of sailboat surface. The boat is old and the surface was untreated. The measured area is a square 10 μm by 10 μm. Since the vertical scale is in nanometers, it is magnified by about a factor of 15 compared to the horizontal scale. (Thanks to CheHwi Chong, Robert Geer, and Stephen Olson for these measurements.)

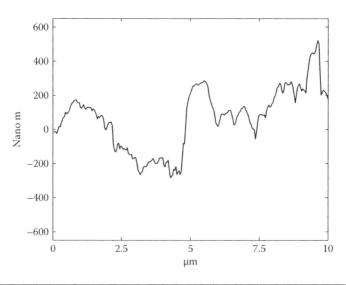

Figure 7.3 A cross-sectional cut obtained from Figure 7.2.

The two roughness profiles shown in the Figures 7.2 and 7.3 differ from the profiles of Figure 7.4 and 7.5 because the samples were prepared differently. The first pair of figures shows the measured roughness of an untreated surface. Since this Flying Scot sailboat was 25 years old, stored outside, and never polished or waxed, these figures represent the roughness of thoroughly weathered gelcoat. This second

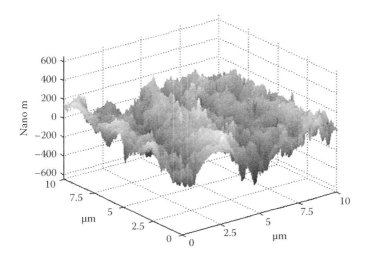

Figure 7.4 A surface characterization from the same boat but for a carefully sanded surface.

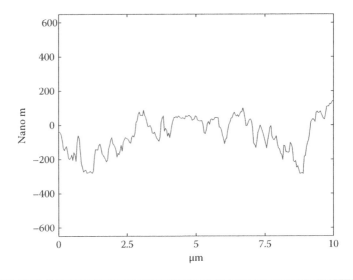

Figure 7.5 A cross-sectional cut from Figure 7.4.

piece, taken from the same sailboat, was subjected to a vigorous sanding regimen, ending with the finest grit sandpaper available (#2000 from an auto body shop). The mean particle diameter in #2000-grit sandpaper is about 5 µm.

One conclusion is clear. It is possible to make an old gelcoat surface smoother. The improvement is about a factor of 2, but the effort was considerable. Even with machinery, it would take a lot of work to make an entire sailboat surface smooth on the scale of #2000-grit sandpaper. Also, some gelcoats have a surface layer impervious to water that is better left intact. The character as well as the magnitude of the roughness is a little different on the sanded surface.

A second conclusion will be clear to some sailors. Smoothing the gelcoat is not worth the effort. Even the untreated and weathered surface is very smooth. To other sailors, a perfectly smooth surface is the Holy Grail of sailing, and no surface is smooth enough. It does not pay to argue with Holy Grail seekers.

These surface profiles were obtained using an "atomic force microscope." This reveals features too small to be seen optically. To obtain the profile, a tiny needle approached the surface until an atom was "felt." Then the position of the probe was recorded. This was done on a square of 512×512 points, scanning like a TV scans its picture. The resulting 262,144 data points were used to generate each of the two-dimensional profiles. The one-dimensional curves are central cross sections of the two-dimensional data.

No data were obtained for a surface with wax or other applied coating. It seems unlikely that wax would make the surface smoother. Although wax may feel slippery because its molecules interact very weakly with the molecules in your finger, this is not relevant for sailboat performance unless the wax keeps small particles from sticking to the hull.

7.2 Inadequate Theory

It is surprising that even basic questions about surface drag do not have simple answers. Sailors ask particularly difficult questions. They want to know the tiny difference between the small drag on a pretty smooth surface and the (possibly) smaller drag on a very smooth surface. Although one can certainly detect the increased drag of a very rough surface, experimental or theoretical comparisons of the tiny differences for nearly perfect surfaces are very difficult. The sticky-surface boundary condition of the Navier–Stokes equation means that even a perfectly smooth surface presents significant viscous drag. It is not clear if additional drag due to roughness will be proportional to the roughness. The term "hydrodynamically smooth" is sometimes used to suggest that roughness has no effect if is it below some minimum level. The comments that follow are a summary of commonly held beliefs. In my opinion, they should be approached skeptically.

In the forward sections of a boat, roughness can instigate turbulence. This is generally bad because turbulence increases the viscous force. A large Reynolds number is the primary signpost of turbulence. *A very smooth fluid flow combined with a smooth surface can significantly forestall the onset of turbulence.* But the wind's flow is surely random and in most cases the water is turbulent as well. The importance of a very smooth surface is not clear when the incident fluid flow is turbulent. The golf ball example presented in Section 7.3.1 suggests that a large amount of roughness can decrease the critical Reynolds number for the onset of turbulence by a factor of 5. However, golf balls are rapidly propelled through the air while sails must deal with the unsteady wind. Thus, the comparison is suspect.

After turbulence is triggered, roughness can still make a difference. A turbulent boundary layer has a "viscous sublayer" next to the surface. Arguments that are less than crystal clear say that *surface roughness within the viscous sublayer should have no effect*. Roughness extending beyond this sublayer extends into the swirling motion of the turbulence and can increase drag. So the problem boils down to estimating the thickness of the viscous sublayer.

As a crude estimate, assume the turbulence has developed and assume the thickness of the viscous sublayer δ does not vary significantly with position. If this is the case, δ can depend only on a fluid speed, the viscosity ζ, and the fluid density. In the sublayer, the relevant fluid speed is not the speed V at a large distance, but v_s, which is a fluid speed in the boundary layer close to the surface. This boundary layer speed is a small fraction of V. The only dimensionally correct expression for the viscous sublayer thickness that depends on the significant physical quantities is

$$\delta \approx \frac{\zeta}{v_s} = (10 \text{ to } 100)\frac{\zeta}{V} \qquad (7.1)$$

If the roughness height is less than δ, the surface is said to be "hydrodynamically smooth" and efforts to make a smoother surface may not be worth the effort. The Fresh Breeze boat speed of 5 m/s gives $\delta \approx 10$ μm for the hull. This is 10 times the roughness of either of the example pieces of Flying Scot gelcoat shown in Figures 7.2 through 7.5.

Air's kinematic viscosity is 15 times larger than water's, so a comparable criterion for hydrodynamic smoothness of a sail is $\delta \approx 0.1 \rightarrow 0.2$ mm. Sails with seams are generally not this smooth, so a new smooth sail should present a little less viscous drag than an old wrinkled sail. If a sail has wrinkles, this could make a difference even though the wind is turbulent.

7.3 Curiosities

A smooth surface is not always the best surface. Some examples of this unorthodox view are worth thinking about.

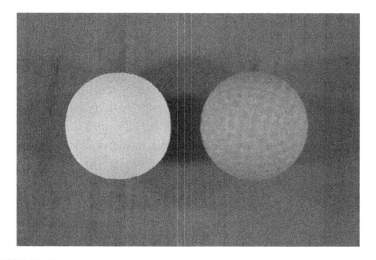

Figure 7.6 Which ball goes further? (Photograph by David Liguori. With permission.)

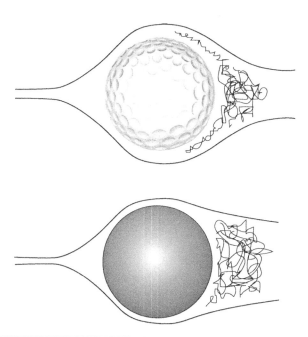

Figure 7.7 For a range of speeds, the dimpled golf ball travels further than its smooth cousin. This figure is a sketch, not the result of a calculation.

7.3.1 Golf Balls

Dimpled golf balls travel farther than smooth golf balls. Has something been missed? Would a sailboat go faster if little holes were drilled into its rudder? It seems unlikely that golfers discovered something unknown to sailors, because sailors are smarter than golfers.

The physical justification of the golf ball dimples is related to the complicated drag crisis mentioned in Section 6.4.2. When a sphere moves rapidly through a fluid, its boundary layer becomes turbulent and more robust. When this happens, the boundary layer does a better job of staying attached to the sphere. The separation point moves downwind, which decreases the size of the low-pressure turbulent region behind the sphere. The drag on the golf ball (shown at right in Figure 7.6) is reduced because the separation point moves downwind, as is sketched for the upper sphere in Figure 7.7. The resulting decrease in drag is the "drag crisis" that occurs at a Reynolds number greater than 2×10^5, as is shown in Figure 7.8. A smooth golf ball is too slow to take advantage of this drag crisis. A solidly hit golf ball's initial velocity is around 60 m/s. The diameter of a golf ball is 4.3 cm. This means the Reynolds number is

$$R(golf) \approx 1.7 \times 10^5 \tag{7.2}$$

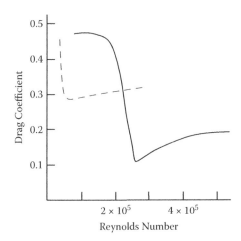

Figure 7.8   The dotted curve shows the decrease in the drag coefficient for a golf ball at intermediate Reynolds numbers. The dark curve is the drag coefficient of a smooth sphere. Both curves are inexact and represent an average of reported data.

Placing 330 to 500 dimples on the ball causes the turbulence and the drag crisis to develop at a speed normally achieved by a tee shot, as shown by the dotted line in Figure 7.8. Golf ball drag is reduced by the dimples for speeds between 18 m/s and 80 m/s. However, for speeds slower than 18 m/s or faster than 80 m/s, the dimples increase the drag. There is still considerable mystery behind this phenomenon. For example, a golf ball with *pimples* does not fly as far as a golf ball with *dimples*.

A similar drag crisis occurs for cylindrical shapes, so one might try the same trick for a mast, ignoring the obvious complications of the attached sail. However, the advantage of putting golf ball dimples on a mast is questionable at best. Wind speeds would often be too small. Then the dimples would increase the drag. Placing dimples on a centerboard is not a good idea. The drag crisis is clearly evident for a sphere and a cylinder, but not for many other shapes.

7.3.2 Swimming Speeds

The maximum human swimming speed is about 2 m/s. Dolphins, sharks, and swordfish are examples of sea animals that can swim faster than 10 m/s. For these predators, success depends on speed. Despite their common dependence on speed for survival, these animals have evolved alternative paths to fast swimming.

Young swordfish have scales. However, the swordfish sheds its scales as an adult. This suggests smoother is faster, but dolphins and sharks have different speed tricks.

Half a century ago, M. O. Kramer concluded that dolphins do not consume enough food to swim as fast as they do for as long as they do. This curious bit of science reminds one of August Magnan's conclusion from 20 years earlier that bumblebees can't fly. There may be as many explanations of dolphin speed as there are dolphins. Some example suggestions are listed here:

- Dolphins have a smooth skin that is soft and pliable, and this suppresses turbulence.
- The dolphin's outer layer can flake off every two hours. The flakes may break up vortex motion.
- Associated with the flaking is the expelling of ethylene oxide, and a thin gas layer should decrease the effective viscosity.

- Some dolphin skins have furrows that are reported to change the vortex motion to reduce the drag.
- The dolphin has skin folds called "dermal ridges" running parallel to the dolphin body. Motion of these ridges is supposed to suppress eddies.
- The bottle-nosed dolphin has quasi-periodic ridges running around the body, not along it. These may reduce the effects of vortex filaments.

All these ideas are untested. It is hard to measure a dolphin's speed. Even if you can get a speed estimate, you can't ask the dolphin if he/she was really trying his/her best, and you can't tell the dolphins to turn off their ethylene oxide to see whether that makes a difference. When data are scarce, theories abound. There are stories that the Japanese (who else?) will construct robot dolphins to test these theories.

Sharks approach the speed problem in a different way. Shark scales, called *denticles*, are tiny (less than 1 mm across) and made of tooth-like material. They stick up from the surface. Most sharks feel smooth when stroked from head to tail, but they feel like sandpaper when stroked in the other direction. (Please stroke your shark carefully.) The individual denticles on fast-moving sharks are very smooth on their leading edges. Farther back, the denticle crown has ridges with depths on the order of half the width. Theories that explain why these denticles could improve speed are not easy to understand.

7.3.3 Shark Imitations

Although sharks are not considered clever, they appear to have invented a reduced drag surface that is worthy of imitation. Microgrooves, or riblets, are a rough analogy to the shark's denticles. They consist of parallel ridges and valleys. Like the shark's denticles, they reduce the viscous force parallel to the grooves. The riblets are typically V-shaped notches smaller than 1/10 mm across. The riblet drag reduction is reported to be as large as 5% to 8%. If this were true, it would be a big difference.

Dennis Connor and the *Stars and Stripes* won the 1987 America's Cup with riblets taped onto the hull. Shortly thereafter, riblets were made illegal on all racing sailboat classes. In practice, riblets on sail-boats are expensive and difficult to maintain.

Riblets have also been used in rowing competitions, and they have been built into the swimsuits of competitive swimmers. Learjets were some of the first commercial airplanes to use riblets. Details of military aircraft surfaces are not advertised.

7.4 When Is It Smooth Enough?

Sailors only interested in enjoying a day on the water need not worry about surface smoothness. Any standard gelcoat or painted surface will not significantly increase the total drag. The drag differences produced by wrinkles in a sail are similarly small.

Racing sailors often have a much less relaxed attitude toward the surfaces of their hulls and sails. Sailboat races that are kilometers long are often won by margins measured in meters. This suggests changes in surface drag of one part per thousand are significant for the most competitive sailors. There is a problem measuring even simple quantities to one part in a thousand. An example is the length of your sailboat. Try measuring it to a (1/10)% accuracy. Since there are no simple experimental tests to compare tiny differences in surface drag, confusion and contradictions are nearly guaranteed.

It is probably not possible to make a sailboat surface resembling dolphin's skin, and the riblet imitation of sharks is illegal. Rather than imitating marine animals, most sailors set aside the possibility that a smooth surface is not always the fastest surface. They settle for making their sailboat hulls as smooth as possible. The vague arguments about the definition of a hydrodynamically smooth surface means sailors have no unambiguous definition of a surface that is smooth enough.

In my opinion, attempts to achieve overall smoothness to better than a 1 μm scale is a waste of time. But larger scale surface imperfections can sometimes be very important. A small amount of ice on the front edge of an airplane wing can produce catastrophic results, probably because the ice induces turbulence. It is likely that the various nicks and dents that appear on almost all sailboat surfaces can also instigate turbulence. Controlled experiments show that the onset of turbulence can be delayed if two conditions are met. First, the initial flow should not be turbulent and, second, the surfaces should not have the nicks and dents. Sailors cannot control the initial turbulence, but eliminating the noticeable imperfections in a sailboat's surface could still make a difference. So after a collision, try to do a careful repair.

8

WAVES AND WAKES

There is no point in pretending that water waves are simple enough to be boring. Water waves exhibit a refreshing variety of properties. They are dispersive, which means different wave lengths travel at different speeds. Their polarization is neither longitudinal nor transverse because the water travels in circular paths. Even the underlying forces that drive water waves are complicated because both gravity and surface tension determine wave frequencies and speeds. Viscosity plays a role, too, and viscous damping is important for shorter waves. Waves can be generated in a variety of ways, and when the wind is the driving mechanism, the situation is particularly confusing. Finally, waves are nonlinear, which means large amplitude waves are much more difficult to describe than gentler waves. A derivation of many of these wave properties is tedious, so an overview of results that sailors may find interesting is accompanied only by sketchy descriptions of how the results come about.

8.1 Wave Shape

The analogue of the "spherical cow approximation" for water waves is the sine wave. Sine waves have the smooth periodic shape shown in Figure 8.1a, with the up and down undulations continuing to the left and right without end. Water waves that are generated by light breezes resemble sine waves.

Waves generated by even moderate winds develop higher and more pointed peaks than sine waves and their troughs become relatively flat. In strong winds, the waves become unstable and "white caps" appear at the wave crest. The highest wave that does not yet produce white caps was first theoretically described by Stokes. It is shown in Figure 8.1b. The peak-to-trough height H of this tallest Stokes wave is

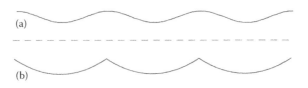

Figure 8.1 (a) The idealized sine wave shape of a low amplitude water wave. (b) The shape of a larger amplitude wave just before it becomes unstable to white cap formation.

only one seventh the wave length λ. (The wave length is the distance from one peak to the next.) The slope of a Stokes wave at its peak is 30° above horizontal. If the wind dies, the waves can last for quite a while, but the sharply pointed Stokes waves change fairly quickly to the sinusoidal shape.

It may seem surprising that a single water wave is never nearly as high as it is long. One must remember that the water's surface is typically covered with a complicated mix of many waves with different wave lengths propagating in different directions. The sum of these waves can occasionally produce anomalously high and steep waves. Also, sailors may exaggerate wave heights because they can be frightening.

The wind-generated waves a sailor experiences never resemble the waves of Figure 8.1. The example in Figure 8.2 shows that it is often hard to recognize either the sine wave or the Stokes wave in the real world.

In addition to their height H and wavelength λ, waves are characterized by their frequency f and (phase) speed v_p. They are related by

$$v_p = \lambda f \qquad (8.1)$$

Equation 8.1 is fairly intuitive. The length of a wave λ (in meters) multiplied by the number of up and down oscillations in one second, f, gives the wave speed in meters per second.

Most of the formulas that follow are strictly valid only for the simple small-height sine waves shown in Figure 8.1a. However, they represent reasonable approximations for all water waves, including the peaked Stokes wave of Figure 8.1b.

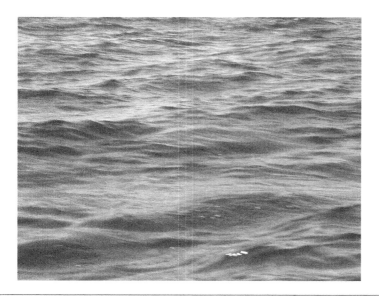

Figure 8.2 Real waves. (Photograph by Sally Snowden. With permission.)

8.2 Water Motion

When a sine wave moves across a body of water, it does not carry the water with it. Instead the water sloshes up and down and back and forth. The path of a drop of water near the surface moves in a circle whose diameter is the wave height, as shown in Figure 8.3. The speed, u, of the water drop is always smaller than the wave speed

$$u = \frac{\pi H}{\lambda} v_p \qquad (8.2)$$

Equation 8.2 follows, because the water must travel a circle of circumference, πH, in the same time it takes the wave to move one wavelength, λ. At the peak of the wave, the water moves with the wave, but the water backs up in the troughs. Between crest and trough, the water and the wave are moving up or down. The complicated mix of horizontal

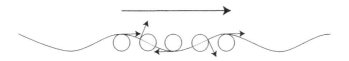

Figure 8.3 The water at the surface of a wave travels in circles. The velocity at each point is indicated by the smaller arrows. The speed of the wave itself is larger, as indicated by the large arrow.

and vertical motion is a consequence of water's incompressibility. When water is falling, nearby water must move away horizontally to make room. Later, water is "sucked in" to a region where the surface is rising.

For waves longer than a boat length, sailors can take advantage of the wave structure when sailing downwind. Just follow two simple rules.

1. Stay on the front slope of the wave, near the top. That way the sailing is both downhill and downstream.
2. Never sail in the wave troughs. Climbing out of a trough involves sailing uphill and against the flow.

Of course, these rules are impossible to follow. Nonetheless, athletic sailors on small boats can significantly improve their speeds by a clever management of the boat position on waves.

The circular motion of the water is largely confined to the surface. The speed with which the water moves at a depth d decreases exponentially as

$$Speed \approx \exp\left(-\frac{2\pi d}{\lambda}\right) \tag{8.3}$$

In other words, at a depth of $\lambda/(2\pi)$, the motion is decreased by almost a factor of 3. For sailboats, more than most boats, this wave motion can be disturbing. Under typical wave conditions, a centerboard or keel often extends to water depths where the wave motion is insignificant. Thus, a wave coming from the side pushes the boat back and forth, but the centerboard is anchored in still water. The result is a torque that tips the boat from side to side, and this torque flops the sail back and forth. This is a good reason to raise the centerboard when sailing sideways to the wind. Of course, leeway and weather helm must also be considered when adjusting the centerboard depth.

The circular motion of the water in a wave is an idealization valid only for very shallow waves. For realistic situations, there is also a drift of the water in the direction the wave moves, but this drift velocity is always very small compared to the wave's phase velocity. To a reasonable approximation

$$v(drift) = 2\left(\frac{\pi H}{2\lambda}\right)^2 v_p \tag{8.4}$$

Since the wave height, H, is always less than a seventh of the wave length λ, applying this formula suggests a drift velocity that is never more than a tenth of the wave velocity. Langmuir in 1938 found a surface current in the water that was about 2% of the wind speed. Some of this motion was no doubt caused by the wind dragging the water, and not just the $v(drift)$ of Equation 8.4. Many more detailed studies of drift have followed Langmuir's early observations, but they are less entertaining.

8.3 Gravity Waves

Gravity is the driving force of waves when the wavelength is greater than a few centimeters. Gravity pulls water from the peaks into the valleys, but the water's momentum carries it too far down, so it gets pushed up again.

8.3.1 Wave Frequency

The back and forth motion of water being pulled down by gravity and pushed up by buoyancy is characterized by a wave frequency

$$f = \sqrt{\frac{g}{2\pi\lambda}} \qquad (8.5)$$

Here $g \cong 9.8 \text{ m/s}^2$ is the acceleration of gravity. Equation 8.5 can be partially justified by simple arguments. One might expect the frequency to depend on all the wave properties: wavelength, λ; height, H; the water density, ρ; and the acceleration of gravity, g. However, the mass density, ρ, cannot affect the frequency for essentially the same reason that all objects fall at the same speed regardless of their mass. Since the frequency formula applies to waves with arbitrarily small height, H should not be in the formula. The only combination of the remaining quantities, g (in meter/second^2) and λ (in meters), which yields the unit of frequency (inverse seconds) is the square root of g/λ. There is a sneaky way to derive the $1/\sqrt{2\pi}$ in Equation 8.5. It is based on the observation that a drop of water balanced on a wave would be accelerated down the slope of wave the same way a skier accelerates downhill. But this drop of water is part of the wave, so the acceleration should be the same as the acceleration of the water in the wave. This observation leads to Equation 8.5 with the appropriate constants.

8.3.2 Wave Speed

Combining Equation 8.1 for wave speed with Equation 8.5 for the frequency shows that the wave speed is proportional to the square root of the wavelength.

$$v_p = \sqrt{\frac{g}{2\pi}} \sqrt{\lambda} \tag{8.6}$$

For a "typical" medium-sized wave with $\lambda = 1$ m

$$v_p(1 \text{ m } wave) \cong \frac{5 \text{ m}}{4 \text{ s}} \tag{8.7}$$

In wind conditions where waves of this size are observed, sailboat speeds are often faster than (5/4) m/s.

Waves can be much longer and move much faster in the ocean. Storms can generate wavelengths of 300 m or more. Using Equation 8.6 again yields

$$v_p(300 \text{ m } wave) \cong 22\frac{\text{m}}{\text{s}} \tag{8.8}$$

This speed is slightly faster than the record for the world's swiftest racehorse. Normal sailboats do not travel as fast.

You can sit on the shore of the ocean and calculate the speed and the wave length of the "rollers" running up on the shore, using nothing more than a watch. The wave period, T, which is the inverse of the frequency, f, is the time interval between successive waves crashing up on the beach. Equation 8.5 gives the wavelength, λ, in terms of $1/f = T$. Then Equation 8.1 shows that dividing the wavelength by the period, T, gives the phase speed, which yields in Equation 8.9 a practical way to calculate the wave speed and length.

$$v_p = \frac{g}{2\pi}T \cong 1.56\frac{\text{m}}{\text{s}^2} \times T$$

$$\lambda = \frac{g}{2\pi}T^2 \cong 1.56\frac{\text{m}}{\text{s}^2} \times T^2 \tag{8.9}$$

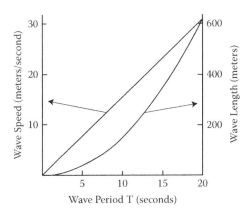

Figure 8.4 Straight line: The speed of a water wave as a function of the wave period, T. Parabola: The wave length of the water wave as function of the wave period, T.

The period-dependence of the wave speed and wavelength is illustrated in Figure 8.4.

An example illustrates this method of characterizing waves. If waves arrive every 15 s, Equation 8.9 or Figure 8.4 informs you that the wave length and speed were $\lambda = 350$ m and $v_p = 23$ m/s. One should be careful when estimating a period. In practice, waves are not perfectly periodic, and they are not all the same size, as can be seen in Figures 8.2 and 8.6. If the period you observe is half a minute or longer, you are probably only counting every other wave. If not, the waves were generated by a really violent storm.

The velocity and wavelength obtained from Equation 8.9 are valid only when the wave is in deep water. As waves enter shallow water, their period is unchanged, but they move more slowly, their wavelength shrinks, their height increases, and they can bend around corners. When the water depth becomes less than the wavelength, Equation 8.6 is changed. The wavelength is replaced by 2π times the depth of the water. This means the speed is determined by the water depth instead of the wavelength, and all waves travel at about the same speed. Tsunamis generated by earthquakes are examples of shallow-water waves even in a deep ocean because their wavelengths are measured in kilometers. Because these waves travel at a constant speed, they do not spread out with time. This makes them much more dangerous when they approach a shore.

8.4 Capillary Waves

Surface tension replaces gravity as the dominant driving force for waves with wavelengths λ significantly less than 1.7 cm (100 cm = 1 m). The surface tension σ characterizes the force needed to stretch the water's surface. The stretching is larger for shorter wavelengths, so the wave speed of these "capillary waves" increases as the wavelength shrinks, which is the opposite of gravity waves. The form of the phase velocity can again be qualitatively justified just by looking at the units.

$$v_p = \sqrt{\frac{2\pi\sigma}{\rho}}\sqrt{\frac{1}{\lambda}} \tag{8.10}$$

The mass density of freshwater is $\rho = 1{,}000$ kg/m^3 and the air–water surface tension is $\sigma \cong 0.073$ N/m at 20 °C. Thus, the phase speed of a capillary wave with a 1½ cm wave length is

$$v_p(1/2 \text{ cm}) \cong 0.3 \frac{\text{m}}{\text{s}} \tag{8.11}$$

When sailors see these short wavelength, capillary waves riding on the top of longer gravity waves, they know it is very windy. It takes a lot of wind to sustain these short waves because they dissipate their energy quite quickly.

At intermediate wavelengths, both gravity and surface tension contribute to the wave velocity. The combination of forces means water waves have a minimum speed of 0.22 m/s when the wavelength is 1.7 cm.

8.5 Damping

Waves don't last forever. They slowly die as their energy is transformed into heat. As described in Section 7.3.4, the kinematic viscosity ζ is the friction-like term of fluid mechanics that produces dissipation. The height of a sine wave decreases exponentially with time, so $H \propto \exp(-t/\tau)$. That means τ is the time needed for the wave amplitude to decrease by nearly a factor of three. The decay time is always a squared distance divided by the kinematic viscosity. For the water wave,

$$\tau = \frac{\lambda^2}{8\pi^2} \frac{1}{\zeta} \tag{8.12}$$

Here $\zeta \cong 10^{-6}$ m^2/s is the kinematic viscosity of water. The $1/(8\pi^2)$ in Equation 8.12 is left as an exercise for the truly ambitious sailor.

In practical units, the water wave damping time of Equation 8.12 is

$$\tau \cong \frac{5}{4} s \times \left(\frac{\lambda}{1 \text{ cm}} \right)^2 \tag{8.13}$$

Thus a 1 cm wave lives only about 5/4 s, and a shorter wave would have an even shorter lifetime.

For longer wavelength gravity waves, the viscous damping is much less important. A 1 m wave has a lifetime of hours. However, the damping time of Equation 8.12 applies only to the sine waves with small height. Energy dissipation is much more aggressive when the wave height H becomes large. It can be clearly seen when waves develop white caps.

A sailor looking for fresh wind keeps wave lifetimes in mind. It is the shorter waves that are an indication of wind. Long waves without little ripples on the top can be remnants of a wind long gone. The physics behind seeing the wind through wave observations is described in Chapter 10.

8.6 Wind and Waves

A simple but old-fashioned idea describes how the power of wind is fed into waves. Once a wave is created it produces a small barrier to the wind. Wind blowing over the water pushes on the wave in the same way it pushes on a sail. As with the sail, the force is roughly proportional to the square of the difference between the wind speed, W, and the wave speed, v_p. Thus, waves can gain energy from the wind as long as the wind is moving faster than the waves. When you observe a sequence of waves with a period $T = 15$ s and use Equation 8.9 to deduce a wave speed of 23 m/s, you can be sure the wind speed that generated these waves was at least 23 m/s.

The power delivered to a wave is proportional to the force on the wave multiplied by the wave speed (just as with sails), yielding

$$Power \propto (W - v_p)^2 v_p \qquad (8.14)$$

The wave speed that maximizes the power in Equation 8.14 is

$$v_p \approx \frac{1}{3} W \qquad (8.15)$$

Typical sailboats move downwind with half the wind speed, so one expects that sailboats can overtake freshly formed waves. This is the typical situation that occurs on relatively small bodies of water and light winds.

When the wind is moderate or strong, it can supply an abundance of energy to the waves of many wavelengths. Although waves with one-third the wind speed may be most quickly created, very short waves will appear because the wind has enough energy to overcome the damping. Longer waves whose speeds approach the speed of the wind can also be energized. Eventually, the faster waves with longer wavelengths will dominate the water landscape because they can have greater height. Thus, on large bodies of water, the speed of the dominant waves is comparable to (but always less than) the speed of the wind that created them.

8.6.1 Flat Water

Very light winds make no waves at all. The surface of the water is like glass. Sailors take this as a sign to drink beer and watch TV (or pick your own leisure activity). Two effects inhibit wave formation in the lightest winds. One is the minimum wave speed of about 0.22 m/s. This slowest possible speed occurs because waves are driven by both gravity and surface tension. Since wind can push a wave only if it is moving faster than a wave, W must be greater than 0.22 m/s before a wave could start. This explanation mostly misses the mark. The waves produced in the lightest breezes usually have wavelengths of several centimeters, significantly larger than the 1.7 cm of the slowest wave. Using Equation 8.13 shows that a 1.7 cm wave has a lifetime τ of only about 3.6 s. Light winds are not capable of refreshing the wave energy in such a short time. On the other hand, the lifetime of a 6-cm wave is three-quarters of a minute. This is a much more easily sustained wavelength. The wind speed W must be at least the 0.3 m/s, and probably double this, to produce the 6-cm waves. The wind speeds near the water's surface are smaller than the wind speeds at the sail, so waves typically appear on water only when $W \approx 1$ m/s.

The variation of wind speed with altitude is quite variable. Sometimes when the water is colder than the air above, the wind is nearly stratified. In such cases, the water surface can indicate calm conditions even when there is significant wind at sail level.

The mechanism by which waves first appear on perfectly flat water surface is mysterious. One might ask how the wind transfers energy to the waves if there is no wave to push against. Part of the answer lies in the turbulent nature of wind. The atmosphere is characterized by fluctuations in both pressure and velocity, and the pressure fluctuations that push the water surface up and down can start wave motion even on a flat surface. In special cases, one can see a boat sailing over flat water with wind ripples following behind. This may be due to the boat's wake stirring up the water so the wind has something to "grab onto." It could also be caused by the sail deflecting a wind at mast height down to the water's surface.

8.6.2 Fetch

Waves are not instantaneously generated. They gradually build up amplitude as they plow downwind. The simplest assumption is that the wave energy per unit area increases at a constant rate. That is,

$$\frac{Wave\ Energy}{Area} \propto Fetch \qquad (8.16)$$

Here "fetch" is the distance to the point where the wave generation starts, often a shore that is directly to windward. The wave energy is proportional to the square of its amplitude. That means the wave height should be proportional to the square root of the fetch. This simple argument is consistent with the phenomenological expression derived mostly from observations

$$H \approx \frac{1}{200} W \sqrt{\frac{Fetch}{g}} \qquad (8.17)$$

In Equation 8.17, the number 1/200 is pretty vague and has no sound theoretical justification. Also this formula applies only over distances where the waves are growing. Because waves cannot move faster than the wind and wave heights cannot exceed about one-seventh of a wavelength, the wave growth eventually stops. Assuming average wave speeds equal one-third the wind speed, W, gives a crude estimate of the distance over which waves will continue to grow

$$Fetch(\text{max}) \approx 50 \frac{s^2}{m} W^2 \qquad (8.18)$$

Thus in a Fresh Breeze with $W = 10$ m/s, the waves continue to grow for about 5 km. For a Gentle Breeze with $W = 5$ m/s, wave heights will continue to grow for only about 1.25 km, and in light air the waves do not go far at all before they are at maximum height. For longer waves whose speeds are closer to the wind speed, the maximum fetch would be larger.

The formulas relating fetch, wave height, and maximum fetch are crude phenomenology. The coefficients above are roughly reasonable for smaller bodies of water where most recreational sailing is done. Different coefficients would better fit the data for ocean waves. The wind generates a broad spectrum of wave heights and wavelengths, and the average wavelength and wave speed also increase with fetch.

Waves make a significant difference in sailboat speed. Sometimes they can help downwind sailing. Waves are always a hindrance for sailing against the wind. By sailing closer to a windward shore and thereby decreasing the fetch and wave height, a sailor can improve upwind sailing speed, provided the wind is not also diminished near the shore. However, since the wave height varies with the square root of the fetch, one must sail four times closer to the shore in order to decrease the wave height by a factor of 2.

8.6.3 Wind and Wave Energies

Anyone who has experienced the buffeting of the ocean surf knows that waves carry a great deal of energy. The energy per unit area of the water's surface is

$$\frac{Wave\ Energy}{Area} = \frac{1}{8}\rho(water)gH^2 \qquad (8.19)$$

Equation 8.19 is obtained from the observation that force times distance is energy. A term ρgH is the weight of a tub of water with a depth H and an area of one square meter. The extra H is associated with the distance the water is lifted. With the appropriate factors of 2, this energy is double the amount of work needed to produce a sinusoidal trough-crest structure. The total energy is twice this work energy because waves have an additional kinetic energy (energy of motion) that is equal to their potential energy.

Wind also carries considerable energy, but it is naturally expressed as energy per unit volume instead of the energy per unit area that characterizes the waves

$$\frac{Wind\ Energy}{Volume} = \frac{1}{2}\rho(air)W^2 \qquad (8.20)$$

These energies can be compared if one considers only the wind energy up to some standard height, D. Because $\rho(water) \cong 800\rho(air)$, the wave energy is quite large. Some rough approximations are needed. Assume the wave speed is half the wind speed, so $v_p = W/2$. This determines the wavelength λ in terms of the wind speed via Equation 8.6. Assume the wave height is one-seventh the wavelength, as is the case for a Stokes wave. Using these approximations gives

$$\frac{Wave\ Energy}{Wind\ Energy\ up\ to\ D} \approx K\frac{W^2}{D} \qquad (8.21)$$

Here, $K \approx 1$ s^{2}m.

With Equation 8.21 one can compare the wave energy with the wind energy up to the top of a mast by taking $D = 10$ m. In a Fresh Breeze with $W = 10$ m/s, the waves have 10 times as much energy as the wind below 10 m. At this wind speed, waves have as much energy as the wind up to 100 m. If one wishes to harness nature's power to obtain "green" energy, it appears that one could gain as much energy from the waves as the wind. Unfortunately, the occasional violent storm at sea makes plans to harness the energy of waves a very challenging engineering problem.

8.7 Wave Packets and Group Velocity

Real waves do not oscillate forever and they do not extend as far as the eye can see. Additional curiosities of waves can be seen when their finite extent is considered. A "wave packet" is a simplified version of a wave confined to a fairly small region of the water's surface. An example is shown in Figure 8.5.

The wave packet moves, but at a different speed than the individual waves peaks. The wave packet speed is called "group velocity" and is denoted v_g. For gravity waves more than a few centimeters long, the group velocity is only half the phase speed (also called *phase velocity*).

$$v_g = \frac{1}{2} v_p \tag{8.22}$$

Anyone subjected to the wake from a speeding power boat is familiar with the difference between the phase and group velocities. The wake often appears as a sequence of a few big waves. Watching the peaks of these waves reveals a relatively high speed. It would appear that the annoying wake will pass quickly. Sadly, it takes a surprisingly long time before the bouncing and rocking stops. Individual waves seem to magically appear at the back end of the wave packet, speed to the front and then disappear. This happens again and again, resulting in a group velocity that is only half the phase velocity. The two-to-one

Figure 8.5 A wave packet. This group of waves moves half as fast as the individual peaks.

ratio of phase velocity to group velocity is the key to understanding the V-shape of the wake that follows behind a boat. (See Section 8.9.)

A wave packet will slowly spread out and individual wave heights in the packet will become smaller. Wave heights decrease because of the speed difference between the long and short wavelengths that compose a wave packet. Roughly speaking, after a wave packet has traveled a distance d, the wave height is

$$H(d) \approx \frac{H_0}{\sqrt{1 + \left(\frac{d}{4w}\right)}} \tag{8.23}$$

Here, H_0 is the initial height of the wave packet, and w is the initial width of the wave packet. This formula suggests a wave packet will travel 12 times its original width for the amplitude to decrease by a factor of 2, but it must travel 60 times its original width before the amplitude is decreased by a factor of 4.

When the initial wave packet can be described by a large initial width w, the denominator of Equation 8.23 stays small for a long time. Waves generated by a storm many kilometers across can be roughly described by a large w. Waves from such a storm can travel long distances before they become harmless. On the other hand, when large waves are produced in a small area, the smaller w means their large height quickly decreases.

There are many reasons why Equation 8.23 is only a rough estimate. Details of the original wave packet shape, the definition of wave packet width, and nonlinear corrections for high waves will all influence the decay of a wave packet. The group velocity of tsunamis and other shallow-water waves is the same as the phase velocity. That means the wave does not spread out and Equation 8.23 does not apply. Tsunamis can travel very large distances while maintaining their height.

For capillary waves, the group velocity is 50% larger than the phase velocity. This is the opposite of gravity waves. Individual wavelets can't keep up, and they disappear at the back end of the wave packet.

8.8 An Example

Real-world waves, such as those shown in Figure 8.2, are not simple. They can be viewed as a continuous onslaught of wave packets or a mixture of waves with many frequencies. They are only roughly periodic. This is illustrated in the wave heights measured 50 km off the coast of Monterrey, California, that are shown in Figure 8.5.

Measurements were made four times a second, which is not rapid enough to see the fine features of wavelets that move on top of the longer length waves. One can see that the waves over this 5-min period are only roughly periodic. A simple counting would suggest about 40 waves in 300 s. This corresponds to a period $T = 7.5$ s, or a frequency of 0.133 Hz. (One Hertz is one cycle per second.) Using Equation 8.9 or Figure 8.4 tells one that waves with a 7.5-s period have a speed of 11.7 m/s, so the wind that produced this wave pattern was blowing at least 11.7 m/s.

The wind speed was measured at the same place and time that the waves were observed. The result is shown in Figure 9.1. Although the wind speed is variable, it never approaches 11.7 m/s. The apparent contradiction comes about because the waves were produced in the past by a stronger wind. Waves last a long time, but the wind can die quickly.

The waves shown in Figure 8.6 can be viewed as a superposition of many perfectly periodic waves. The transformation of Figure 8.6 into its frequency components (via a Fourier series) gives additional information

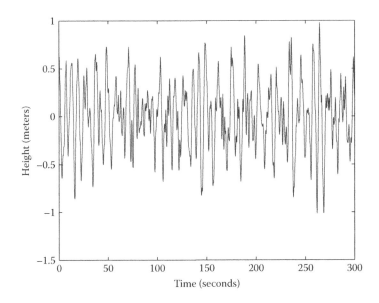

Figure 8.6 Five minutes of wave heights in the Pacific Ocean measured four times per second. (Thanks to Scott Miller for wave and wind data.)

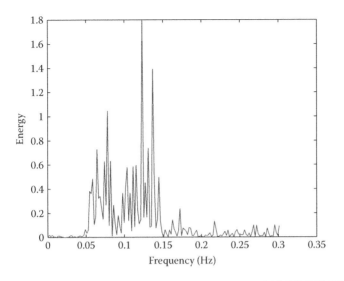

Figure 8.7 The frequency distribution of the energy of the wave pattern shown in Figure 8.6.

about the wave structure. The energy (or squared Fourier amplitude) stored in each frequency is shown in Figure 8.7. This figure shows that the maximum wave energy does correspond to a frequency of about 0.133 Hz. However, the wave spectrum is spread out, and there appears to be a second peak at half the 0.133 Hz frequency. One could argue that these lower frequencies correspond to waves with twice the period and thus, through Equation 8.9, double the velocity of 11.7 m/s. This is doubtful. Nonlinear effects could lead to an alternating of big and small waves and this effect is not described by the simple theory. However, it is safe to say that the peak at 0.133 Hz is real, and the waves shown in Figure 8.5 were generated by a wind speed that was at least 11.7 m/s.

8.9 Wakes

If you quickly move your finger across the water's surface, the wave pattern that follows your finger is a tiny version of a wake. It is essentially the same wake produced by a duck, a sailboat, or the Queen Mary. All wakes have three pieces in common a center portion that follows directly behind the boat and two side arms making a V shape. There may be smaller waves between the center and side wakes. Outside the V there is no wake.

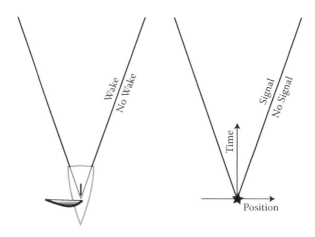

Figure 8.8   The wake's V shape and special relativity's light cone.

Because it has this sharp outer boundary, a wake is like the "light cone" of special relativity. One boat can feel the wake of another boat only if it lies within the V of the other boat's wake. From relativity, we know that if something happens at position $x = 0$ and time $t = 0$, it can cause something else to happen at another position x' and a later time t' only if light traveling from the first event arrives in time to cause the second event, which means $|x'| \le ct'$. Here c is the speed of light. Graphing this relation between x' and t', gives the characteristic V of the light cone that resembles the V of the wake. A comparison is shown in Figure 8.8. A fundamental concept of relativity is the constancy of the speed of light and the angle of the light cone. The light cone has the same shape no matter how fast an observer moves. Wakes have a vaguely analogous nifty property. No matter how fast a boat is moving, the angle of its V is the same. Albert Einstein loved sailing. Who knows the true source of his inspiration?

8.9.1 Properties

A summary of wake properties and geometry are presented here. A derivation of the results, which is an annoying geometry and trigonometry exercise, is postponed to Section 8.9.3.

8.9.1.1 Center Wake   The waves in the center wake move with the boat, so they have a phase velocity $v_p(center)$ equal to the boat speed U.

Using Equation 8.6 for a wave's phase velocity,

$$v_p(center) = \sqrt{\frac{g\lambda(center)}{2\pi}} = U \qquad (8.25)$$

Here, $\lambda(center)$ is the wavelength of the central wake. Equivalently,

$$\lambda(center) = \frac{2\pi}{g}U^2 \qquad (8.26)$$

That means you can determine the speed of a boat by observing the separations between wave crests in the wake that follows along behind. A Fresh Breeze boat speed $U = 5$ m/s means $\lambda(center) \cong 16$ m. Doubling the speed produces a wake four times as long, so the center wake wavelength is an accurate way to characterize boat speed. Streamlined objects, such a ducks, canoes, and some sailboats produce only a faint central wake.

8.9.1.2 Side Wakes   Boats of any size moving at any speed greater than about 1 m/s produce a side wake with universal properties. The triangle shown in Figure 8.9 characterizes the common geometry of all wakes.

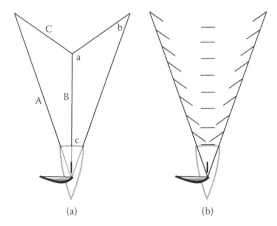

(a)   (b)

Figure 8.9   (a) The basic wake triangle, with sides A, B, and C shown on one side and the corresponding opposite angles a, b, and c shown on the other side. (b) The lines representing wave peaks of the central and the two side wakes. The side wake peaks are parallel to the line C shown at left.

Side A of the triangle lies on one of the V-lines of the wake. Side B follows directly behind the boat. Side C is parallel to the individual wave crests that make up the wake. The relative lengths of the triangle sides are

$$A = \sqrt{6}$$

$$B = \sqrt{3}$$

$$C = 1 \tag{8.27}$$

The corresponding angles a, b, c opposite these sides are

$$\sin(c) = \frac{1}{3}; \quad \text{or} \quad c \cong 19.5^0$$

$$b = \frac{1}{2}(90^0 - c) \cong 35.25^0$$

$$a = 180^0 - b - c \cong 125.25^0 \tag{8.28}$$

Also, $\sin(a) = \sqrt{2/3}$, $\sin(b) = \sqrt{1/3}$.

Individual wave crests in the side wakes do not have the same phase speed as the boat. The phase velocity of each wave is equal to the component of the boat's speed parallel to the wave motion. That means

$$v_p(side) = U \sin(a) = U\sqrt{\frac{2}{3}} \tag{8.29}$$

The relation between phase velocity and period (Equation 8.9) means one can deduce the speed of a boat from the period T of the side wake it produces.

$$U = \sqrt{\frac{3}{2}\frac{g}{2\pi}}T = 1.91\frac{\text{m}}{\text{s}^2}T \tag{8.30}$$

Thus, if a wake bounces a stationary observer up and down once every 6 s, the boat that made that wake was moving fairly fast; $U = 11.5$ m/s. This relation is valid whenever the boat producing the wake is moving in a straight line at a constant speed.

The viscous damping of most wakes is negligible because the wave lengths are typically on the order of 1 m or longer. Wakes eventually die out because they form a wave packet. That means dispersion (wave spreading) rather than damping is the mechanism that decreases the height of the wake waves. The approximate wave height $H(d)$ after it has traveled a distance d is given by Equation 8.24, repeated here.

$$H(d) \approx \frac{H_0}{\sqrt{1 + \left(\dfrac{d}{4w}\right)}} \tag{8.24}$$

As before, w is the original size of the wake train. As with other wave packets, the fairly rapid initial decrease in height becomes quite slow as the wake widens. Energy conservation means the width of the wake (and the number of waves in the wake) increases with time. After some time a wake changes from a few very high waves to many gentler waves.

Sometimes wakes appear to last forever, and a sailboat may find itself riding a powerboat's wake that never seems to stop. This may not be an illusion. The simple theory described here is based on the assumption of gentle low-height waves. Some wakes, being quite steep, may exhibit "solitary wave" characteristics. Idealized solitary waves have shapes that do not decay. Instead, they propagate intact—essentially forever. The suggestion that wakes can be solitary waves has met considerable and justifiable skepticism. Despite this, it is possible that nonlinear effects may extend a wake's lifetime.

8.9.2 Wake Energy and Hull Speed

The power delivered to a wake means extra force is needed to keep the boat moving. Heavy sailboat wakes become the major drag at higher speeds. The "hull speed" is essentially the maximum speed of a heavy sailboat in any reasonable wind. It is traditionally determined by postulating that wake resistance overwhelms all other forces when a wake wavelength becomes comparable to the boat length. Using Equation 8.6, that means

$$Hull\ Speed \approx \sqrt{\frac{g}{2\pi}\ Length} \tag{8.31}$$

This suggests that a boat 8 m long would have trouble traveling faster than 3.5 m/s. In practice, this limit is too restrictive and the hull speed estimate of Equation 8.31 is often multiplied by a number somewhat larger than unity.

It is impossible to design a wake-free boat, but lighter sailboats that can skim over the water (plane) produce much shallower wakes. Sailboats that plane are not subjected to the hull speed limit.

Although the hull speed limitation is one of the most familiar nautical concepts, a simple explanation is illusive. Computer calculations and sophisticated tank tests allow professionals to understand and predict wakes, but nonprofessionals appreciate simple explanations. Two traditional, complementary, and unsatisfying models are used to justify the hull speed formula. They are the "two wakes merge to one" picture and the "sailing uphill" picture. They are followed by a nontraditional scaling model that I find more satisfying.

8.9.2.1 Two Wakes Merge to One   For large boats, one can often see two wakes generated; one is produced by the bow pushing water up and to the side. A second wake appears at the stern as water is suddenly allowed to uplift. At hull speed, these wakes are supposed to merge to give an especially large wake that requires an especially large power to maintain. However, the effect is not spectacular. When two wakes of height H are independent, the energy is proportional to $2H^2$. (The factor of 2 is from the two wakes.) If the two wakes combine to make a single wake of height $2H$, the energy is proportional to $(2H)^2 = 4H^2$, so the energy is only doubled. When a boat speed is increased past the point where the wave amplitudes add, the simplest argument would say the wake drag should return to lower values.

8.9.2.2 Sailing Uphill   A sailboat rides on its own wake. At high speed, the wake's wavelength becomes comparable to the boat length. The water piles up at the bow and leaves a hole at the stern. The boat is constantly trying to climb out of the hole. However, at even higher speed, the wake wavelength becomes longer than twice the boat length. When this happens riding up the broader hill should become easier.

These arguments suggest that hull speed is more like a sound barrier, which could be overcome. Heavy boats never overcome this barrier.

8.9.2.3 Scaling Model   An alternative model of wakes is described here. Although this model does not explain why hull speed should scale with the square root of the boat length, it does show how wake drag can increase very rapidly with boat speed.

The basic assumption is that the wake produced by a duck is essentially the same as the wake of the Queen Mary. Only the distance scale is changed. This assumption allows a comparison of the wake force at different speeds. One can compare the wake of a boat moving at two different speeds. From this, one can obtain ratios of the wake energies. The energy ratio gives a power ratio. Since power is force times speed, this gives the force ratio.

Consider two boats, traveling at different speeds, $U(fast)$ and $U(slow)$. Assume the wavelength of the faster boat's wake is twice that of the slower boat. The relation between wavelength and speed (Equation 8.6) means $U(fast) = \sqrt{2} \cdot U(slow)$. Let $E(fast)$ and $E(slow)$ be the total energies contained in the wakes of these two boats up to a distance equal to N wavelengths behind the boat. Because the two wake patterns are assumed to be scale models of each other,

$$\frac{E(fast)}{E(slow)} = 4\left(\frac{H(fast)}{H(slow)}\right)^2 \qquad (8.32)$$

The factor four appears because the N wavelengths of the faster boat wake is twice as long and twice as wide as the corresponding wake section of the slower boat. The quantities H in this expression can be taken as the height of the largest wake wave. The squared ratio of these heights appears because wave energies are proportional to the square of the wave height (Equation 8.19). Assuming the same scaling applies to the wave height, $H(fast)/H(slow) = 2$ means

$$\frac{E(fast)}{E(slow)} = 16 \qquad (8.33)$$

The power needed to produce a wake is the wake energy divided by the times it takes to generate the wake. The faster boat takes a time that is greater by $\sqrt{2}$ because the time is obtained from the doubled distance divided by a speed that is larger by $\sqrt{2}$.

$$\frac{Power(fast)}{Power(slow)} = \frac{16}{\sqrt{2}} \qquad (8.34)$$

Since power is force time velocity, and the velocity ratio is $\sqrt{2}$, the ratio of drag forces is obtained by dividing by another $\sqrt{2}$

$$\frac{F_D(fast)}{F_D(slow)} = 8$$

This means increasing the speed by $\sqrt{2}$ (a 41% speed change) increases the wake drag force by a factor of 8.

The speed ratio does not have to be $\sqrt{2}$ for the scaling approach to apply. Since $8 = (\sqrt{2})^6$, the scaling really says the wake drag force should be proportional to the sixth power of the speed.

$$F_D(wake) \propto U^6 \qquad (8.35)$$

A force proportional to the sixth power of the speed is negligible at low speed but it becomes overwhelming at high speed.

The total drag force is then the sum of the conventional drag, which is proportional to the square of the speed, and a wake contribution proportional to the sixth power of the speed.

$$F_D = \frac{1}{2}C_D(U^2 + BU^6) \qquad (8.36)$$

A graph of the variation of drag force with speed is shown in Figure 8.10. Because the coefficient B in Equation 8.36 is not determined by a scaling argument, the scales on Figure 8.10 are undetermined.

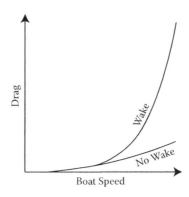

Figure 8.10 The drag force without wake is proportional to the square of the boat speed. The scaling model adds a drag proportional to the sixth power of the speed.

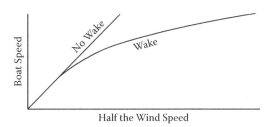

Figure 8.11   An estimate of downwind boat speed as a function of wind speed. The linear dependence of boat speed on wind speed changes to the lower curve when wake drag is added.

With this model of wake drag, the hull speed corresponds to the speed at which the BU^6 term cannot be overcome by any reasonable wind. However, the boat speed limit is not really sharp. A calculation of the downwind boat speed as a function of the wind speed is shown in Figure 8.11. This result is a generalization of the downwind boat speed obtained in Section 2.3.2. The water's drag including the wake is set equal to the wind force. For simplicity, the downwind speed factor S_0 was taken to be unity. The straight line in Figure 8.11 ignores the wake and reproduces the results of Chapter 2.

The U^6 form of wake drag is not appropriate for lighter boats. As their speed increases they partially lift from the water. This produces a smaller wake so the scaling argument does not apply.

Dolphins know that wake generation produces a large drag force at high speeds. Being mammals, they must breathe air. Being fast swimmers, they do not want to generate a wake. To accomplish these conflicting desires, they swim underwater. When they must surface to breathe, they come up only briefly and sometimes leave the water entirely.

8.9.3 Wake Properties Derived

Intuitively, one would think a wake would narrow as boat speed increases. This is not the case. The following explains why the wake angle c shown in Figure 8.9 is always given by $\sin(c) = 1/3$.

Two geometric constraints on the wake establish its properties. The first requires that the wake (as a whole) moves with the boat. The second requires that individual wave peaks also move with the boat. These two requirements are illustrated in Figure 8.12

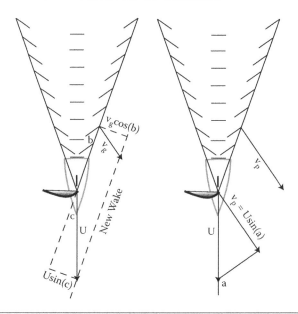

Figure 8.12   The wake as a whole and the individual wake peaks must keep up with the boat.

In one unit of time, the boat moves south a distance U. The wake must keep up. In the direction perpendicular to the outer edge of the wake, the distance moved is $U \sin(c)$, not U. The wake moves at the group velocity in the direction perpendicular to the individual wave peaks, which make and angle b with the wake's outer edge. So the outer edge moves a distance $v_g \cos(b)$. Equating these two measures of the wake's motion and using the relation between group and phase speed ($v_g = v_p/2$ from Equation 8.22) gives

$$\frac{1}{2} v_p \cos(b) = U \sin(c) \qquad (8.37)$$

In this same unit of time, an individual wave in the wake moves a distance v_p. The boat moves with this wave, partly by sliding sideways and partly by moving in the same direction. One can equate the wave's phase speed to the component of the boat's velocity that is parallel to the motion of the individual wave. This gives

$$v_p = U \sin(a) \qquad (8.38)$$

The next part is the trigonometry. Combining Equations 8.37 and 8.38 allows one to eliminate the speeds.

$$\sin(a)\cos(b) = 2\sin(c) \tag{8.39}$$

The sum of the three interior angles of a triangle is π radians. Thus,

$$\sin(a) = \sin(\pi - b - c) = \sin(b + c) \tag{8.40}$$

so

$$\sin(b + c)\cos(b) = 2\sin(c) \tag{8.41}$$

The obscure trigonometric identity $2\sin(b + c)\cos(b) = \sin(2b + c) + \sin(c)$ means Equation 8.41 can also be written as

$$\sin(2b + c) = 3\sin(c) \tag{8.42}$$

This result does not yet solve the problem. It appears that the angle c can be anything, depending on the value of b. However, there is a trick. The largest possible value of c corresponds to the outer edge of the V-shape of the wake. Since $\sin(2b + c)$ can never be greater than unity, one obtains an the upper limit on c.

The universal wake shape is determined by the upper limit on the sine of the wake angle.

$$\sin(c) = \frac{1}{3} \tag{8.43}$$

This famous result for the outer edge of the wake was first obtained by Lord Kelvin in 1887. The original derivation is quite complicated. Once c is determined, the other wake properties follow relatively easily.

The wake is strongest at its outer edge because many values of b give almost the same value of c when $\sin(c) \cong 1/3$. At its maximum angle, many different mini-waves add to make the whole. This effect is a little like throwing a ball straight up into the air. The ball spends an especially long time near the top of its path because its speed vanishes at the maximum height.

8.10 The Importance of Waves

The formal characterization of waves by frequency, wavelength, and height almost belittles their significance. Ocean waves can be enormous and a real danger to safety. Faced with a wave of prodigious height, the sailor will probably not waste time calculating the wave's energy—even though it would be interesting to know. On smaller bodies of water, dingy sailors may not have to deal with a wave "as big as a house." But even on small lakes, there are plenty of submerging bows and curious wave-assisted capsizes on windy days.

Boat designs cannot ignore waves. Waves influence the stability and speed of sailboats, and sailboats on a beat must constantly cut through waves. There is a reason most sailboats are pointed at the front end. Much of the stress on hull structure comes from the incessant pounding of the waves. Without waves, sailboats construction would be quite different. The boats could be light and agile if they did not have to withstand repeated buffeting.

The wakes produced by sailboats are also more than a curiosity. An efficiently designed boat must not produce a large wake. This is a serious design problem since scale models of boats cannot properly replicate wake effects. Since wake drag can become dominant at moderate to high speeds, hull designs with hydrofoils or other options which help the boat lift from the water are a key to high-speed sailing. A boat that moves well in still water but is jerked about by moderate waves is not acceptable. A boat that moves well in light winds but produces a large wake at high speed is not acceptable.

But waves are not all bad. Sometimes a sailboat can surf on a wave and achieve remarkable speed. Sailors "see" the wind by the waves it produces. Sailboat strategy would lose one of its most important tools if waves did not tell the story of the wind.

Every day of sailing offers the sailor a different set of waves. Sometimes they are intimidating and sometimes they are useful. Waves are never boring.

9

WIND

Three vaguely defined and overlapping effects determine the wind. They are turbulence, weather, and geography. Roughly speaking, turbulence rules wind variations over relatively short times and distances, while weather determines average winds. Even more roughly speaking, weather can be predicted and turbulence cannot. Geographical features like shorelines and hills modify the winds produced by weather and turbulence.

9.1 Two Examples

The best way to grasp the confusing nature of the wind is to look at real examples. The first example is the "steady wind" shown in Figure 9.1. This wind was measured simultaneously with the water-wave measurements shown in Figure 8.6. The wind velocity was recorded four times a second, at a height of 9 m above sea level, at a fixed point in the Pacific Ocean. The average wind speed of about 7.5 m/s is the "weather" part of the steady wind. The roughly ±1 m/s fluctuations about this average are the turbulence

The second example is the "gusty breeze" shown in Figure 9.2. This wind is shown for a much longer time, about 7.5 h. The gusty breeze data comprise more than 100,000 wind speed measurements taken four times a second. The data are so compressed that fine details appear only as a blur. Because the gusty breeze was measured 10 m above land, it is more characteristic of sailing conditions on small bodies of water where land effects typically produce more variation. The slight decrease in the gusty breeze over the period of a day and the significant decrease in fluctuations toward the end of the day are associated with weather. The fluctuations on the shortest time scale are turbulence. It is hard to say if other features in Figure 9.2 represent predictable structure (weather) or

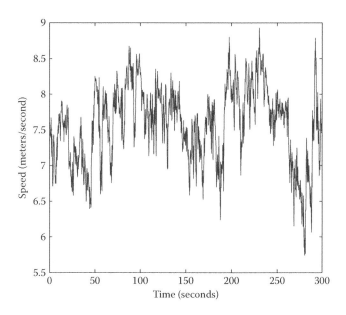

Figure 9.1 The horizontal wind speed of the "steady wind" recorded four times a second over the Pacific Ocean.

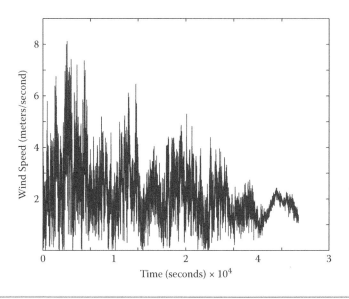

Figure 9.2 The horizontal wind speed of the "gusty breeze" measured four times a second for 7.5 h.

only randomness (turbulence). Some may see periodicity in this wind pattern and others may not.

The quarter-second wind velocity data were obtained by measuring the speed of sound pulses sent between closely placed sources and receptors. Sound travels faster when it is moving with the wind, so if one carefully measures the time it takes for a sound pulse to pass from a source to a receptor, one obtains the wind speed in that direction. To gain a full picture of the wind speed and direction, three differently aligned source-receptor pairs were used. The data in Figures 9.1 and 9.2 are the horizontal part of the wind speed. Even though this technology shows wind variations on an unusually short time scale, the measurements are still too infrequent to reveal the quickest fluctuations produced by the wind's turbulence. In a Fresh Breeze, wind fluctuations occur over times that are much less than a second, perhaps milliseconds.

9.2 Turbulence

Turbulence is hard to define and harder to quantify, but it is easy to recognize. Turbulence is unpredictable, almost by definition. It produces the ever-present minute-by-minute and second-by-second wind variations that make sailors' lives so interesting. It is wind's nature to be always changing, never reproducing and never typical. For sailors, this means "set it and forget it" is never a correct way to sail. Efficient sailing requires endless vigilance and constant adjustment. Sailors sometimes find turbulence frustrating, but it has its romantic aspects. Turbulence makes the stars twinkle.

9.2.1 Details of the Gusty Breeze

Turbulence is especially noticeable in the gusty breeze example shown in Figure 9.2. The turbulence associated with the steady wind of Figure 9.1 is similar in character, but less dramatic. In Figures 9.3, 9.4, and 9.5, the first 34 min of the gusty breeze have been expanded. These figures show how the wind direction varies, how even an averaged wind speed varies, and how the wind blows vertically as well as horizontally.

Fluctuations in wind direction always accompany fluctuations in wind speed. The wind speed and direction are compared for the first half hour of the gusty breeze in Figure 9.3.

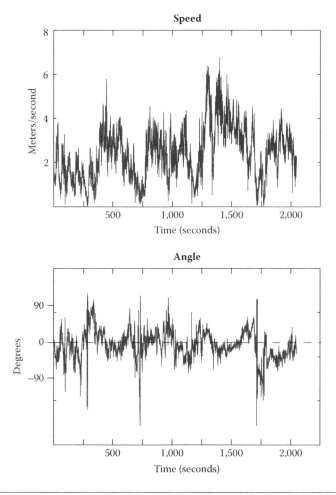

Figure 9.3   The first 34 min of the gusty breeze (shown in Figure 9.2). Both wind speed and direction fluctuate. The angle spikes associated with vanishing wind should be ignored.

The negative angle spikes in Figure 9.3 are numerical anomalies that should be ignored. They occur because the wind direction is not defined when it vanishes.

An attempt to make sense of turbulent wind is frustrating. For example, a periodicity seems to be poking through the noise in Figure 9.3. But if one examines the next half hour of the data shown in Figure 9.2, this particular hint of periodicity is gone. If there is a correlation between wind speed and direction, Figure 9.3 does not make it obvious.

Humans have an ability to see patterns and periodicities in situations where none exists. Gambling casinos and astrologers are fond of these people. Even though statistical evidence for a periodic wind is meager, many expert sailors claim to discover periodicity and other complex patterns in the wind. Because these experts do well in sailing races, it is probably best not to argue with experience.

There is little theoretical reason to expect wind to be periodic, but periodicity is not absolutely impossible. For example, a "Karman street" is a periodic precursor of fully developed turbulence in which vortices (whirlpools) of opposing circulation are produced. Karman streets are really a characteristic of essentially two-dimensional fluid flow. Observations of Karman streets in the atmosphere have been reported when stratified wind flows through mountain passes. There may be other special situations that produce periodic oscillations, but such situations should be very rare.

One can observe periodicity in many phenomena that involve the fluid flow of either the wind or the water. However, this is often the result of a coupling of the fluid motion to a mechanical system. Without this interaction, the turbulent fluid would probably not appear to be periodic. For example, one can sometimes feel a periodic oscillation by briskly dragging the handle of a canoe paddle though the water. The periodic sound vibrations associated with speaking, singing, and snoring are produced by a coupling of air's motion and our vocal cords. Sometimes high-speed sailboats make a humming sound. These oscillations may be associated with a "shedding" of vortices with alternate circulations, but in most cases, the mechanical coupling is a key to the periodicity.

Sailors are not quick enough to deal with quarter-second wind variations. Also, the quarter-second fluctuations take place on distance scales shorter than the length of a sailboat, so sailors at the bow and stern experience different winds on the shortest time scales. Sailors should respond to wind variations that last on the order of a minute. An average of the measurements of Figure 9.3 eliminating the rapid fluctuations is shown in Figure 9.4. A similar smoothing occurs when the angle is averaged.

Most sailors are unaware of the vertical part of the wind velocity. This up–down motion is unavoidable. A fast wind often climbs over or dives under a region of slow air that lies in its path. This means the

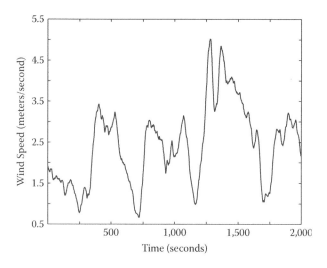

Figure 9.4 A smoothing of the gusty breeze speeds from Figure 9.3.

wind blows at an angle ϕ above or below the horizontal. The observed variation of this angle for the same half hour of gusty breeze shown in Figures 9.3 and 9.4 is shown in Figure 9.5. It is surprising that this angle can exceed 45°, which means the vertical part of the wind is occasionally stronger than the horizontal part. However, the rapid oscillations about zero mean the time-averaged vertical component of

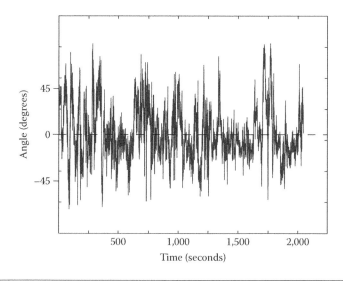

Figure 9.5 The angle above or below horizontal at 10 m above the surface for the gusty breeze.

the wind is nearly zero. At an altitude of 10 m, the up–down part of the wind must be short-lived because it has to stop before it reaches the surface. At the height of a couple of meters where a typical sailboat crew sits, the vertical wind is less significant. Higher up, where eagles fly, the vertical wind is even more important and longer lasting.

The data of Figures 9.1–9.5 show only the time variation of the wind. The spatial variation of the wind is another aspect of turbulence that is equally important for sailors. Unfortunately, graphs of the spatial dependence of the wind are not available, since this would require an enormous number of wind detectors.

Even though no precise measures exist, the spatial nonuniformity of wind is apparent to sailors. They can see wind variations on the surface of the water. These wind variations generally travel downwind with roughly the average wind speed. Thus, the wind of the future tends to be the wind to windward. Since downwind sailing is typically only half the wind speed, the wind of the future for downwind sailing is best judged by looking behind the boat. When sailing downwind, you usually cannot catch the wind that lies ahead. Although wind fluctuations often follow the average wind for a while, they are still transitory. Turbulence is not simply the steady drifting of wind patterns across the water's surface. Sailors are often disappointed when an approaching puff of wind suddenly disappears. They are surprised when a wind gust appears without warning.

9.2.2 Turbulence Theory

Turbulence results from the instability of fluids with high Reynolds numbers. The Reynolds number $R = W \cdot L/\zeta$ is a product of characteristic velocity and distance divided by the (kinematic) viscosity, ζ. For air, $S \cong 1.5 \times 10^{-5}$ m²/s, typical wind speeds are meters per second and the distance (the height of the turbulent atmosphere) is hundreds of meters or more. Except in a virtual calm, this means the atmosphere's Reynolds number greatly exceeds one million, which clearly signals turbulent instability.

One needs more than a Reynolds number to describe atmospheric turbulence. Different weather conditions can alter the nature and extent of the added turbulence. For example, when warm air lies below cold air, the atmosphere is vigorously mixed because the less dense warm air tries to rise. The opposite case of warm air overriding cold air produces relative stability against vertical mixing. Since turbulence is driven by instability, its extent is strongly influenced by temperature profiles.

The backbone of turbulence theory was elegantly developed by Kolmogorov (and others), presented in 1941, refined over the years, and finally withdrawn (in part) in the 1960s. Despite the remarkable success of this theory in a wide range of applications, Kolmogorov concluded that the intermittent character of turbulence makes it even more complicated than the theory outlined here.

Some of Kolmogorov's ideas are related to the "Richardson Cascade" (1929), nicely summarized by his rhyme:

> Big whirls have little whirls which feed on their velocity.
> Little whirls have lesser whirls, and so on to viscosity...

Curiously, the Richardson cascade was really Ezekiel's idea.

> The wheels had the sparkling appearance of chrysolite, and all four of them looked the same: they were constructed as though one wheel were within another. They could move in any of the four directions they faced, with veering as they moved.

Today, Ezekiel's wheels and Richardson's whirls are called "eddies." Less poetically,

1. Kinetic energy in the form of large-scale motion is supplied to the atmosphere through the sun's heat and the earth's rotation.
2. This energy eventually dissipates through viscous heating.
3. The heating can only take place through motion on a very small distance scale.
4. The character of turbulence is largely determined by the way the energy flows from the large-scale to the small-scale motion.

Energy is dissipated by a "bucket brigade" transferring energy density from the largest scale down to the smallest scale where viscosity can do its work, changing the kinetic energy to heat. But the buckets don't contain water. They are filled with energy density. And the energy flows not from place to place, but from eddy to eddy, each of a smaller size. Crudely, an eddy is a circulating motion with a characteristic size λ.

The transfer of energy to smaller eddies does not take place through energy diffusion. Instead, a large eddy breaks up into smaller eddies, and this process is repeated many times over. It is a bit like Mickey Mouse as the Sorcerer's Apprentice, where the bucket-carrying brooms appear to subdivide without end.

The kinetic energy density in the atmosphere has units Joules/kilogram. The rate at which the energy density flows to smaller scales is called ε. It has the units of energy density per second, or a velocity cubed divided by a distance.

Consider first the energy density flow ε from the largest eddies. This number should depend on the largest eddy size l, the average extra wind speed Δu, and the fluid density ρ. There are no other relevant physical variables. The only expression one can construct from these variables that has the right units is

$$\varepsilon \approx \frac{J}{Kg \ s} = \frac{m^2}{s^3} \approx \frac{(\Delta u)^3}{l} \qquad (9.1)$$

Solving for the eddy velocity gives

$$\Delta u \approx (\varepsilon l)^{1/3} \qquad (9.2)$$

The same idea applies to the smaller eddies in the bucket brigade. To avoid energy pile-up, energy density must flow through these at the same rate, so they are characterized by the same ε. (In a bucket brigade, everyone must work at the same speed.) The analogous dimensional argument applies as before, except the characteristic length is now the size of the smaller eddy. Thus, the eddy velocity on the scale of λ is

$$\delta u_\lambda \approx (\varepsilon \lambda)^{1/3} \qquad (9.3)$$

The wind-speed measurements shown in Figures 9.1–9.6 were taken at a fixed point, but the wind was moving past at an average wind speed, W. Thus, measuring after a time interval τ is almost the same as measuring at different points separated by a distance $\lambda = \tau W$, so the time-dependence of wind fluctuations are described roughly by

$$\delta u(\tau) \approx (\varepsilon \tau W)^{1/3} \qquad (9.4)$$

Equation 9.4 quantifies the observation that the wind is always fluctuating. If you measure the wind at $t = 0$, and measure it again at $t = \tau$, the two wind speeds will be different and that difference will typically be proportional to the cube root of the time difference. The cube root function increases very rapidly for short times. This produces the jagged appearance of the winds shown in Figures 9.1–9.5. A more quantitative comparison and validation of the theory is obtained from an examination of the frequency dependence of the energy stored in the wind. The theory predicts wind fluctuations of all frequencies, with a gradual decrease as frequency increases. This conclusion has been verified in a number of experiments.

With this physical picture, the reason the wind speed and direction varies on all time scales makes some sense. The atmosphere is using the only available mechanism to change its kinetic energy into heat. There is a shortest time scale for which the theory applies. For times much less than the quarter-second intervals shown in the Figures 9.1–9.5, the velocity will vary more smoothly. This smoother motion appears when distance scales are so short that the viscosity becomes dominant and motion of air resembles the flow of honey.

9.3 Wind up High

The wind at the top of a mast is stronger than the wind on the deck of the boat. That means all the previous observations relating the true wind $\vec{W}$, the apparent wind $\vec{V}$, and the boat velocity $\vec{U}$ are only approximate because the height was ignored. The wind–altitude story is based on an elegant theory. The results of that theory and its derivation are summarized here. Unfortunately, the real world of sailing does not live up to the assumptions of the elegant theory, so all the conclusions are subject to exceptions and revisions.

9.3.1 Results

Let's start with the theory's weak point. There is a poorly defined minimum height z_1 at which one can assume the wind is essentially zero. Experiments and some intuitive physical arguments show that z_1 is generally quite small. It increases slowly with wind speed and surface roughness (including wave height). A reasonable guess for a Fresh Breeze over water is $z_1 = 1/10$ cm.

Once a guess for z_1 is made, one only needs to plug in the wind speed at 10 m above the surface $W(10\,\text{m})$ and the wind speed at all other heights $W(z)$ are give by the theory. Example results are

$$W(1\text{ m}) = \frac{3}{4}W(10\text{ m})$$

$$W(10\text{ cm}) = \frac{2}{4}W(10\text{ m}) \qquad (9.5)$$

$$W(1\text{ cm}) = \frac{1}{4}W(10\text{ m})$$

$$W(z_1 = 0.1\text{ cm}) = 0$$

There is a rule used to generate these numbers. Every time the height is decreased by a factor of 10, the wind speed decreases by a fixed amount until finally at z_1 the wind speed is zero. For the example of Equation 9.1, one-quarter of the wind speed must be subtracted for each division by 10. If z_1 is chosen to be 1 cm instead of 0.1 cm, the wind-speed variation per decade would be 33% of $W(10\,\text{m})$ instead of 25% of $W(10\,\text{m})$. For both choices, the wind profile is determined by the condition that the wind vanishes at z_1. More formally, this model says the wind speed varies with the log (logarithm) of the height. This model fails at very low height where $z \cong z_1$.

For most sailboats, the ratio of mast height to the distance of the boom from the water is less than 10 to 1. Since z_1 is quite small, this means the difference between the wind at the bottom and top of the sail should generally be less than 25%. This result applies to big and small sailboats alike, and is illustrated in Figure 9.6, which shows the wind variation described by Equation 9.5.

The above assertions about wind speed must be approached with caution. The result is based on the assumption that air of uniform temperature is moving over a fairly flat surface of the same temperature. If cold air moves over warm water, the system is less stable. Denser air above will fall, resulting in increased mixing, which will reduce the change in wind speed with height. When the air is warmer than the water, the atmosphere is more stable. The flow will be effectively layered, and wind speed variations with height can be quite a bit larger.

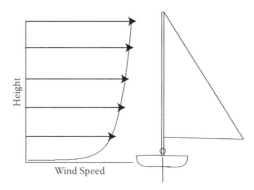

Figure 9.6   The idealized dependence of wind speed with height. Atmospheric conditions can significantly alter this result.

The stratification of wind in one case and the increased mixing in the other are sketched in Figure 9.7. The different temperature profiles are important for sailors. In spring, when the water is cold and the air is warm, an increased wind speed aloft should be apparent. The daily variations in the relative temperature of the air and the water means the wind's altitude profile can vary by the hour.

One can argue that a "twist" in a sail shape can take advantage of the variation of wind speed with height. Since the altitude dependence of wind speed is so uncertain, this justification of twist should also be approached with caution. It seems likely that some twist is desirable even if the wind speed were independent of height. The "Tight Leech" example sail of Section 5.2.1 has no twist, but it is probably not the most efficient sail, even for a perfectly uniform wind.

9.3.2 Theory

The theoretical background of the height dependence of the wind bears some similarities to the general theory of turbulence described

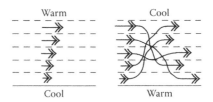

Figure 9.7   The variation of temperature with height can increase or decrease the mixing of high- and low-altitude winds.

in Section 9.2.2. As with turbulence theory, Kolmogorov made important contributions.

Take an arbitrary height above the surface, say 10 m. Air above 10 m is moving faster than air below 10 m (on average). Because air motion is turbulent some of the more rapidly moving air above the 10 m surface will spill into the lower levels. However, in a steady state the wind speed should not increase in time. The average wind speed stays constant because another surface at 9 m is spilling motion into even lower levels. Ultimately, the motion and energy feeds down to the surface where it is absorbed. In the same sense that everyone in a bucket brigade has to work at the same speed, the rate at which wind speed is fed to lower levels cannot depend on the height z. A quantitative measure of this transfer of wind to lower levels is called v_*^2. Assume extra wind speed flows through an area, *Area*, for a time t into a volume, *Vol*. Then the mean speed in that volume will increase by an amount ΔW given by

$$Vol \cdot (\Delta W) = v_*^2 \cdot Area \cdot t \qquad (9.6)$$

It follows that v_* has the units of a velocity.

The variation of the mean wind speed with height is essentially determined by v_* and z because there are no other relevant physical quantities. The change in speed with height $dW(z)/dz$ must be written in terms of these quantities, and it must have the units of speed divided by distance. There is only one combination of v_* and z that can accomplish this. An examination of units alone requires that

$$\frac{dW(z)}{dz} = bv_* \frac{1}{z} \qquad (9.7)$$

where b is an unknown dimensionless constant.

This last equation is the justification for the assertion that the speed depends on the log of the height because the log of z is the only function whose derivative is proportional to $1/z$. If the speed depends on the log of z, its description must follow the prescription of Equation 9.5. There is little point in filling in the final steps needed to determine z_1 because it depends on the unknown constant b.

A key to the results obtained above was the argument that the only physical quantities that can determine $dW(z)/dz$ are v_* and z. This assumption fails as soon as additional variables are introduced. So temperature differences invalidate the theory and one must rely on experiments and more qualitative arguments. There are additional problems, not mentioned here, which make one a little uneasy with the logarithmic dependence of wind speed on height.

9.4 Weather

9.4.1 Predictions and Guesses

Weather forecasts are inexact, especially for wind prediction, but they do allow sailors to make educated guesses. Alert sailors can also look at the sky and make intelligent judgments. It takes a keen eye and a lot of experience to relate cloud formations to wind developments. Sometimes cloud movement or "cloud streets" predict the wind direction, but sometimes they don't. Sometimes approaching clouds mean more wind, but sometimes less.

Electronic media update local weather conditions on a minute-by-minute schedule. This has largely supplanted the old-fashioned methods of weather prediction. Most sailors forget to consult their barometers before heading out to sea.

Everyone has heard the saying;

"Red sky at night, sailor's delight.
Red sky in the morning, sailors take warning"

But few sailors get up early to check the sky color.
Some poetic lore is true more often than not.

"Winds of the daytime wrestle and fight
Longer and stronger than those of the night"

But there are exceptions.

"If it rises at night
It will fall at daylight."

The basic point is just a reiteration of what everyone knows. You can't rely on the weather. Despite this, the simplest aspects of weather described here are worth keeping in mind.

9.4.2 High-Pressure Systems

The sun heats the atmosphere unevenly. Heated air rises and leaves low pressure behind. Low pressure in the Northern Hemisphere produces counterclockwise circulation. Falling cold air produces a high-pressure region and clockwise circulation in the Northern Hemisphere. The first descriptions of the rotational motion are sometimes attributed to Heinrich Wilhelm Dove (1803–1879), but some aspects of wind circulation may have been understood by Aristotle's school.

The circular motion of high- and low-pressure systems is caused by Earth's rotation. It is easiest to visualize the circulation at the North Pole, even though this is not a preferred sailing venue. If there is a high-pressure region at the North Pole, the winds initially fan out in all directions. At the North Pole, "all directions" are south. From the viewpoint of an observer fixed on the North Star, the air is pushed out along straight lines, and the Earth is rotating in a counterclockwise direction underneath the wind. From the viewpoint of someone standing on Earth, the Earth is static but the air is spiraling out in a clockwise direction. Both are right because perceived motion depends on the observer's coordinates. These two views are compared in Figure 9.8. The curvature of the path in an Earth-based system can be described in terms of a "Coriolis force," which is really a pseudo-force. This force is proportional to the wind speed.

The clockwise spiral of wind from the North Pole is a simple example of a more general result. The earth's rotation means straight line motion in the Northern Hemisphere appears to turn to the right. This conclusion applies for winds traveling north, east, south, or west. They all turn to the right, which is the reason all high-pressure systems (in the Northern Hemisphere) produce clockwise circulation.

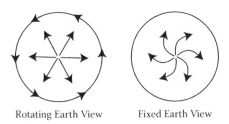

Rotating Earth View   Fixed Earth View

Figure 9.8   If wind travels a straight line when viewed in an absolutely fixed coordinate system at left, the same wind will appear to bend clockwise in coordinates fixed to the Earth at right.

In practice, the circular motion associated with a high-pressure system grows to dominate the radial motion. High-altitude winds in a high-pressure system are nearly aligned with the curves that show constant pressure. More rapid variations of pressure produce higher winds. The typical high-pressure systems of interest to sailors are many kilometers across, and they last for days. During their lifetime these pressure systems are dragged along the Earth's surface by the winds of the upper atmosphere.

Viscosity and interactions with the ground or water slow the near-surface winds. As shown in Figure 9.6, the wind vanishes at the surface. The slowest wind very near the surface does not care about the earth's rotation because the Coriolis force vanishes at zero velocity. The motion of the near-surface air is radial because it is simply pushed out by the pressure. The winds of interest to sailors are an intermediate case, being neither the purely circular high-altitude winds nor purely radial surface creep.

Two idealized high-pressure systems are shown in Figure 9.9. Each side of the figure is an oversimplified weather map. The letters "H" identify the center of high-pressure systems. The circles surrounding the H are curves of constant pressure. As one moves out from the center in any direction the pressure smoothly decreases. The single arrows on the constant pressure curves indicate the clockwise circulation. This circular motion along the constant pressure curves occurs only at altitudes much higher than sailboat masts. The double-pointed arrows denote the wind at a height of a few meters. This is the

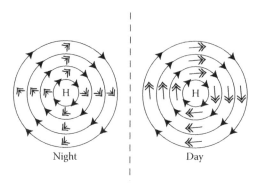

Night   Day

Figure 9.9 The "Night" and "Day" views of an idealized high pressure system. The simple arrows are the high altitude wind. The doubled arrows are the wind at lower altitude.

partially radial and partially circular motion of interest to sailors. One is supposed to notice that "night" differs from "day" in Figure 9.9. The night wind is slower and less circular. The sun's heating and mixing of atmospheric levels is responsible for the difference between night and day. At night, the surface is often cool. This produces a stratified wind flow, as is shown on the left of Figure 9.7. The stratification means the high-altitude wind is cut off from the surface. The day curve assumes the surface is relatively warm. This produces mixing so both the direction and magnitude of the upper atmosphere's motion is more thoroughly mixed with the surface wind.

If weather were really this simple, one could predict the wind. The wind should be light in the morning and increase later in the day when the high-altitude wind is stirred down to the surface. In addition, the wind direction should rotate in a clockwise direction as the sun heats the surface (in the Northern Hemisphere). Weather is seldom that simple. However, on the average, this makes sense. For example, measurements of wind averaged over an entire year in Oklahoma City, Oklahoma, showed that the wind at 4 p.m. was about 11 m/s, but after sunset, the wind dropped to about 6 m/s. The 365-d averaging is the key to this result, since any one day could show very different results. A comparison of the wind up high and near the surface (again in Oklahoma City) showed very small direction difference during the day, but a large difference at night. This observation is consistent with the Figure 9.9.

9.4.3 Low Pressure and Complications

Low-pressure systems are not simply high-pressure systems run in reverse. If there is low pressure at the North Pole (an unlikely occurrence), air would be pulled in from all directions. Again, it turns to the right so air initially headed for the Pole ends up traveling east instead of north. Viewed from above, the circulation is counterclockwise. The circular motion means the air fails to arrive at the center (the North Pole) where the pressure is lowest. The pressure difference and the wind deflection fight against each other. Because the pressure difference is not easily relieved, the pressures and wind speeds can be greater.

There is a second way that geometry maintains low-pressure systems. When something moves in a circle, it tries to fly out from its circular path. This can be described in terms of the "centrifugal force," which is another pseudo-force. Thus, the air circling a low-pressure region has an additional reason resist the attraction of low pressure. A hurricane is an extreme example of a low-pressure system getting out of control. Of course, there is much more to that story.

The air at the center of a low-pressure system rises up and cools, while the center of a high has falling and warming air. Since cold air can hold less water, one expects clouds and precipitations from low pressure. High pressure with falling and warming air produces clear weather.

For a person (or penguin) standing at the South Pole, the winds turn to the left instead of to the right. At the equator, winds turn neither left nor right. At latitude 45° north, the magnitude of the turning is multiplied by $\sin 45° = 1/\sqrt{2}$. One can understand this by interpreting Earth's rotation as a vector $\vec{\omega}$ pointing toward the North Star. The magnitude of $\vec{\omega}$ is the rotation rate and its direction is the rotation axis. At the North Pole, $\vec{\omega}$ is straight up. At the South Pole, it is straight down. At an arbitrary latitude θ, $\omega \sin(\theta)$ is the part of $\vec{\omega}$ that is aimed up. This is the part that is important for rotation. The part of $\vec{\omega}$ that is parallel to the surface is not important.

Low-pressure air rises because its lower density is attracted less by gravity. Warm air is less dense than cold and the natural tendency of warm air to rise and cold air to fall is familiar to anyone living in a drafty house. For our atmosphere, the expansion of air modifies this result, so there is more to the mixing than is indicated in Figure 9.7. Often warmer air at the surface does not rise because it is not warm enough. When air rises, its pressure decreases, it expands, and this expansion means the air is doing work on its surrounding. The energy needed to do the work is taken from the air's heat energy, so expanding air cools. Air will rise 1 m only if it remains warmer than the air 1 m above it after it has cooled.

There is a complication within this complication. When air cools, it can hold less water. If the water condenses, the temperature change is not the same. Just as it takes heat to boil water, heat is given off when water condenses. The condensation contributes to a vertical instability of the air that is sometimes sufficient to generate storms.

Water condensation is also important when warm and cold air masses meet. In this confrontation, the warm air is pushed above the more dense cold air. Rising (and cooling) warm air often has the potential

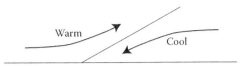

Figure 9.10 Warm air rises over cold air because its density is lower. As the rising warm air cools, it can bring rain.

to produce water condensation and rain. The shape of the warm–cold interface sketched in Figure 9.10 depends on the relative motion the two air masses.

9.4.4 Geography

Shores and hills can guide the wind, changing its magnitude and direction. Sometimes the effect is pretty obvious. A large hill directly to windward will shield sailboats and decrease the wind. In many other cases, the influence of geography is not as predictable as one might hope. Sometimes shorelines appear to channel the wind so it moves more parallel to a shore, but other times the wind appears to be diverted away from a shore. Despite this ambiguity, skilled sailors can often use shorelines and other geographical features to advantage by picking the appropriate course.

A "sea breeze" often develops near the boundary between land and water. The sea breeze is driven by a difference between the temperature of the land and the water. On a sunny day, the land heats more rapidly because the absorption of solar energy takes place in a thin layer. For water, the sunlight's energy is distributed though several meters of nearly transparent water. In addition, water has a very high heat capacity, so a lot of heat is needed to change the water temperature. The air over heated land is warmed by the land. Warm air is less dense so it rises, and the denser air over the water falls and moves in to replace it. The result is a circulation similar to that shown in Figure 9.11. Sea breezes can extend kilometers, so the sailboat in Figure 9.11 is not drawn to scale. When the sun goes down, the land cools more rapidly, and the sea breeze disappears.

The Earth's rotation affects sea breezes like any other wind. Thus, the wind direction is rotated to the right. The extent of the rotation

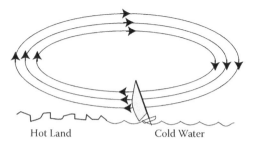

Hot Land   Cold Water

Figure 9.11   A hot land surface heats the air above it, causing it to rise. The air will fall over a cooler water surface. The resulting circulation is a sea breeze. The sailboat is not drawn to scale.

is limited by the geometry of the thermal instability, which generates the sea breeze in the first place.

See breezes can be quite reliable. Other connections between weather and geography can be much more hit-and-miss.

9.5 Apologies

Nearly everything said here about wind and weather is a gross simplification. Wind's turbulence can vary enormously, so the results shown in the first five figures cannot be called "typical." The theory of turbulence presented here is not the last word even for well-controlled situations. For the real outdoor world of weather, there are so many exceptions to the simple ideas that the theory only samples a small bit of reality. Similar reservations apply to the theory of the altitude dependence of the wind. The neglected variation in temperature distribution means that the curve in Figure 9.6 can grossly overestimate or underestimate the steepness of the wind profile. The oversimplifications of the weather comments are obvious. There has never been a high-pressure system as round and symmetric as the one shown in Figure 9.9.

In principle, wind and weather in general can be described by basic physics. In practice, the number of quantities determining the weather can be overwhelming, even though they are essentially for accuracy. I apologize for skipping so much and for violating Einstein's famous rule: "Make everything as simple as possible, but not simpler."

10

STRATEGY

A stock market analogy illustrates the Sailor's Dilemma. A steady east wind is like an uneventful stock market. Steady east winds and uneventful markets are both rare. The market falls ("goes south"). Its analogy is a wind shift to the south. Should you buy or sell? If you are sailing to the east, should you sail on starboard (northeast) or port (southeast)? In both cases, the answer depends on the nature of the change. If the lower stock price and the south wind are random events, you should buy stock and sail northeast, being confident that the market will return to normal and wind will return to the east. But if the market's fall is the beginning of a trend, and the south wind is the beginning of a permanent shift, the correct strategy is completely different. Sell your stock and sail to the southeast when you expect more of the same. This analogy explains why all good sailors are rich.

The Sailor's Dilemma is illustrated in Figure 10.1. If the wind returns to the original east direction, the sailor at top right will be ahead. If the shift to the south increases, the sailor at bottom left will be in the lead.

10.1 Directions

Geometry and directions are important for all aspects of sailing. Sailors should always be aware of the wind direction and its relation to the orientation of the boat. This relative orientation is so important that different names are given different sailing directions. Downwind sailing is called a "run." The wind is from behind the boat, or at least close to behind the boat. A "jibe" occurs when the sail switches from one side of the boat to the other while sailing downwind. A boat is on starboard tack when its sail is on the port side, and vice versa.

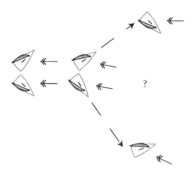

Figure 10.1 At left are the upwind sailing directions in an east wind. At center, a wind shift to the south changes the sailing directions. At right above is the sailing direction if the wind direction returns to the east. Below is the sailing direction if the wind shifts further to the south.

Starboard is the right hand side of the boat when facing the bow. It remains the starboard side even if you turn around and face stern. It remains the starboard side even if your boat is going backward. When the wind is coming from roughly the side it is called a "reach." Sailing upwind is a "beat." A boat that is sailing as close to upwind as is practical is "close hauled." Switching between starboard and port and tack on a beat is a "tack."

10.1.1 Ideal Sailing Direction

The sailor in a race and the sailor eager to make it home for dinner are faced with the same basic question. Which is the best direction to sail? One direction may lead to a shorter path and another direction may produce greater speed, so the choice is not obvious. The "ideal sailing direction" is the best direction to sail for a minimum trip time. If the angle between the Ideal Sailing Direction and the source of the true wind is w, then $U(w)$ is the speed determined by the sailboat's speed diagram described in Chapter 3. The geometry that characterizes the ideal sailing strategy repeatedly uses the notation and speed diagrams described in Chapter 3.

For strategy considerations, the speed $U(w)$ is more useful than the speed ratio $U(w)/W$. Speed diagrams $U(w)$ show the increased boat speed produced by stronger winds. The speed diagrams can be changed into distance diagrams. Multiplying a speed by a time ΔT gives the distance traveled in that time.

10.1.2 Preferred Direction

If a sailor could instantaneously move his or her boat by one boat length in any direction, the direction that would save the most time is the "Preferred Direction." It plays a fundamental role in the formal description of the ideal sailing strategy. It is parallel to a vector $\vec{p}$. As is shown in the following figures, the Preferred Direction often differs from the Ideal Sailing Direction. Sometimes the Preferred Direction is as simple as the direction to the final destination. This would be the case for a reach in a steady wind. When the destination is close to upwind, the Preferred Direction is often directly to windward. When a racing sailboat is near a finish line, the Preferred Direction is in the direction of the finish line. In other cases, the Preferred Direction is quite difficult to determine. The most successful sailor may be the sailor who can best discern the Preferred Direction. In other words, you can't sail well unless you know where you want to go.

*10.1.3 Relation between the Ideal Sailing Direction
        and the Preferred Direction*

Given a speed diagram and a Preferred Direction, geometry determines the Ideal Sailing Direction. To find the Ideal Sailing Direction, draw a line perpendicular to the Preferred Direction (and also perpendicular to the vector $\vec{p}$). Move the line so it just touches the outer edge of the speed diagram. The boat should sail to the point of contact, as shown in Figure 10.2.

In general, whenever two sailboats are sailing near each other (but not stealing each other's wind), the advantaged boat has progressed

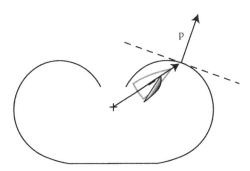

Figure 10.2   The speed diagram touches a line perpendicular to the Preferred Direction. The Ideal Sailing Direction is toward the contact point.

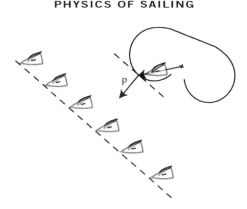

Figure 10.3 Six boats racing toward the southwest have made identical progress in the preferred direction, so no boat is in the lead.

farther in the Preferred Direction. That means being "ahead" in the traditional sense is not always best. This confusing geometry makes sailing a lousy spectator sport for NASCAR aficionados. Despite appearances to the contrary, the boat at the upper left in Figure 10.3 is not in the lead. The speed diagram at the right shows that it is a dead heat because all six boats have made equal progress in the Preferred Direction. Similarly, the iceboat at left in Figure 10.4 is not being given an unfair advantage. All the iceboats are equally advanced in the Preferred Direction.

In special cases, there are two Ideal Sailing Directions. This happens when the line perpendicular to the Preferred Direction touches

Figure 10.4 Iceboats lined up for a "running start." (Photograph by Stéphane Caron. With permission.)

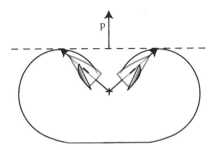

Figure 10.5 When the Preferred Direction is directly to windward, there are two close-hauled Ideal Sailing Directions.

the speed diagram at two places. An upwind example is shown in Figure 10.5. For this case, the two Ideal Sailing Directions correspond to sailing "close hauled." These are the two sailing directions that make the most rapid progress to windward. For close-hauled sailing, the angle between the true wind and the sailing direction is usually around 45°. Sailing at a smaller wind angle is "pinching." Pinching is never the best way to make progress.

10.2 Constant Preferred Direction

Sailing strategy is simpler when the Preferred Direction is fixed.

10.2.1 Condition for a Constant Preferred Direction

Assume the Preferred Direction is directly north. Then the Preferred Direction will not change if the wind does not vary in the east–west direction.

This means the Preferred Direction will remain north if

a. The wind is constant.
b. The wind depends on time but not position.
c. The wind varies only in the north–south direction.

Of course, there is nothing special about north in this characterization of a constant Preferred Direction.

A formal justification of the conditions for a constant Preferred Direction is part of Equation 10.6. The condition makes some intuitive sense. If wind isn't any different "over there," meaning to the east or the west (and you want to go north), there is no reason to change the Preferred Direction.

Once the Preferred Direction is obtained, the Ideal Sailing Direction is determined by geometry. Examples of a fixed Preferred Direction are considered first. Situations where the Preferred Direction can change are described in Section 10.3.

10.2.2 Finish Line

Although it is usually difficult to determine the Preferred Direction, there are some special "easy" cases. One example occurs in a sailboat race very near a finish line. The Preferred Direction is toward the finish line. This is the direction a boat would like to be translated because the boat nearest the finish line is clearly ahead. The construction for this case is shown in the Figure 10.6. The Ideal Sailing Direction is not directed to the nearest point on the finish line because boat speed is increased by sailing at larger wind angle, w. Even if the finish line is at the end of a beat to windward, the sailor in the position shown in Figure 10.6 should not sail close hauled because the finish line is not perpendicular to the wind direction. This is not the only example where sailing close hauled on a beat is not the best strategy.

10.2.3 Upwind in a Constant Wind

An unchanging wind is very unlikely, but it is sometimes a useful approximation. If one wishes to sail upwind in an unchanging wind, the Preferred Direction is fixed and pointed directly to windward.

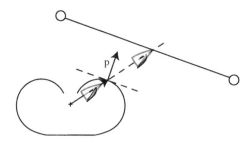

Figure 10.6 At the end of a race, the Preferred Direction points to the nearest point on the finish line. The Ideal Sailing Direction is the quickest path to the finish (the dotted line), but not the close-hauled direction or the shortest distance to the finish line.

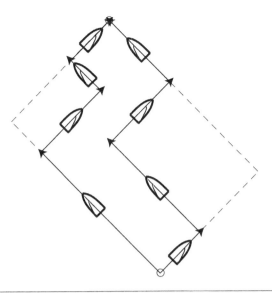

Figure 10.7   Two example paths to windward that take equal time in a steady wind.

This is the only Preferred Direction that allows the boat to sail on both starboard and port tack, as is shown in Figure 10.5. For this special case, the path to windward is not unique. Neglecting the time needed to tack, any combination of starboard and port tacks that brings the boat to the final goal takes an equal time. Two example paths are shown in Figure 10.7. No matter which zigzag path is taken, the Ideal Sailing Direction is close hauled on the steady wind beat. The beat strategy can be much more complicated when the wind is not steady.

10.2.4 Downwind in a Constant Wind

Another simple example is sailing downwind in an unchanging wind. For this case, the choice of path depends on the speed of the sailboat. For slow boats with a downwind speed ratio, $S_0 < 1$, the single straight line path is most efficient. For faster boats with $S_0 > 1$, a zigzag path analogous to upwind tacking is faster. The example in Figure 10.8 compares the correct strategy for two boats with very different speeds. The fast boat has the speed diagram constructed in Chapter 3 for $S_0 = 3/2$. The slow boat is characterized by $S_0 = 2/3$.

   Boat speeds are not really proportional to the wind speed. Often the ratio $U(w)/W$ decreases when the wind increases because waves,

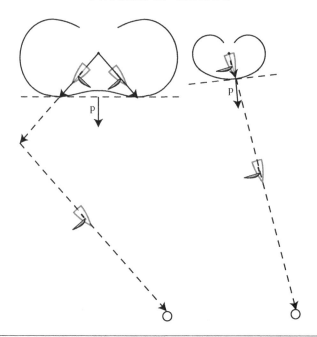

Figure 10.8 The faster boat should jibe downwind. The slow boat should take the boring path and aim directly to its goal.

wakes, and overpowered sails do not allow the boat to maintain a speed proportional to the wind speed. This was shown in Figure 2.10 for real sailboats. Thus, for some sailboats, $S_0 > 1$ when the wind is light, but $S_0 < 1$ when it is blowing hard. When this is the case, it makes sense to jibe downwind when the wind is light and take a straight line course in a heavy wind.

Spinnakers increase downwind boat speed (assuming my favorite trick of dropping one in the water is avoided). This changes the shape of the downwind part of the speed diagram, but it does not change the basic ideas. This may mean a sailboat should tack downwind only when its spinnaker is deployed.

Downwind sailing for the boat with $S_0 = 1$ whose speed diagram is shown in Figure 10.6 (and other figures) is a special case. Because the downwind progress of an $S_0 = 1$ sailboat is constant for a fairly wide range of wind angles, the sailor is free to choose a variety of paths with essentially no time penalty. As is sometimes observed in sailboat races, a fleet of boats sailing downwind will fan out over a fairly large area and converge again at a downwind mark. If the wind is steady,

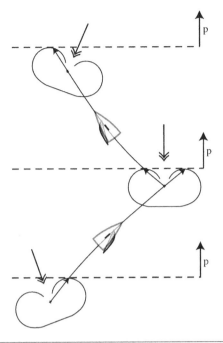

Figure 10.9 The Preferred Direction remains north even if the wind changes in time (or from north to south). The Ideal Sailing Direction and the correct time to change from port to starboard tack are determined by the changing wind direction.

arrival times at the mark make it apparent that no path was particularly advantaged. Although the perfectly "flat bottom" of the $S_0 = 1$ speed diagram is an idealization, it is a reasonable approximation for many sailboats.

10.2.5 Upwind in a Changing Wind

The next two examples describe the slightly more complicated case where the Preferred Direction is either north or south and the wind varies from north to south, or it varies with time. The windward example is shown in Figure 10.9. A sailboat's path is accompanied by its speed diagram at three different positions. The fixed Preferred Direction tells the sailor what path to take and when to tack from port to starboard as the wind evolves toward the east.

In the example shown in Figure 10.9, the sailor should constantly maximize speed to the north, which is not the same as maximizing speed to windward (sailing close hauled) because windward is not always north. When the wind has an east or west component, this sailor should trim sails less tightly and aim a little away from the wind. Close-hauled sailing is appropriate only when the Preferred Direction is directly to windward.

Even in real-life sailing where sailors have better things to worry about than the abstract Preferred Direction, this result has a practical application. Sailors should pick the tack favored by the wind shift. In addition, in a favorable shift, they should sail at a little larger angle to the wind than close hauled.

In practice, one often does not know the Preferred Direction, and one must rely on a best guess. If one expects only random fluctuations about an average north wind, assume the Preferred Direction on the beat to windward is due north. It's a good guess if you have no other knowledge of the wind.

10.2.6 Downwind in a Changing Wind

A similar strategy applies for downwind sailing, as is illustrated in Figure 10.10. The sailor picks a course for maximum speed to the south, not maximum downwind speed. The advantage of jibing downwind is less significant for a slower sailboat whose course should be more directly downwind.

10.3 Variable Preferred Direction

The Preferred Direction is usually not constant because east–west wind variations are common when traveling north or south. The examples of Section 10.2 are seldom an adequate description of real sailing. For the more general case, deducing the least-time sailing path is much more difficult. The side trip through tree rings illustrates one of the ideas which will be generalized to the case of a fleet of identical sailboats.

10.3.1 Rings

When a tree is sawed off cleanly, one can see a sequence of rings, one for each year of growth. Counting the rings gives the age of the tree, with the inner ring being the sapling. Good and bad years mean the tree rings are separated by different thicknesses. The environment

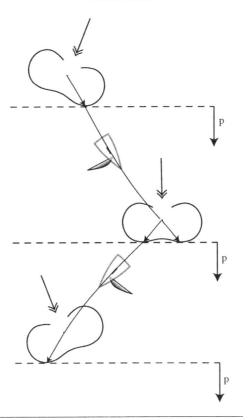

Figure 10.10   When the Preferred Direction is constant but the wind direction changes, the boat should jibe.

may allow different sides of a tree to grow at different rates. This distorts the rings so they are sometimes far from circular, as is the case for the tree section shown in Figure 10.11.

Tree rings are easy to see. It takes some imagination to visualize sailboat rings. If a large number of sailboats were to sail in every direction from a central point, a short time later these boats would form a ring around the starting point. Tree trunks have a natural way of highlighting 1-year intervals to mark their age. For sailboats, one could (in principle) take a sequence of aerial photographs of the boats, and label each photo by the elapsed time T. A superposition of these photos taken at regular intervals would resemble a tree-ring pattern. Large ring separations represent high speed, and narrowly spaced rings are an indication of calm winds. Since winds can be different in different places, the rings could be strangely distorted. If sailboat rings

Figure 10.11 Tree rings. (Photograph © H. D. Grissino. With permission.)

had the shape of the tree rings in Figure 10.10, one would conclude that the wind was blowing harder in the east. The smaller separations of rings toward the outside are analogous to a dying wind.

10.3.2 Sailboat Ring Growth

The sailboat rings are clearly defined only if the sailboats generating the rings are identical and only if each sailboat travels as efficiently as possible toward the next ring. When this condition is met, the sailors automatically generate their own rings. They also generate the Preferred Direction and the Ideal Sailing Direction. This is illustrated by the enlargement of a small piece of a hypothetical sailboat ring structure in Figure 10.12. In the expansion, two adjacent ring segments, corresponding to times T and $T + \Delta T$, are the dotted lines. Three speed diagrams are placed on the inner ring. The speed diagrams are scaled by the small time interval ΔT so they show the possible positions that the three boats could reach in time ΔT. The end boats are sailing efficiently to the outer ring, but the boat in the middle is aimed the wrong way. Because it fails to heed the prescription for the Ideal Sailing Direction, it will lag behind the others. Figure 10.12 shows that the Preferred Direction must be perpendicular to the ring.

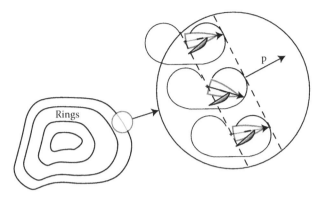

Figure 10.12   A set of nested sailboat rings and an expanded view of one section. The expansion shows how subsequent rings are generated and how the Preferred Direction is determined. The center boat has failed to aim in the Ideal Sailing Direction.

The construction in Figure 10.12 assumed all three sailboats were sailing in the same wind. This spatial uniformity means the Preferred Direction parallel to $\vec{p}$ does not change as time evolves. (The outer ring has the same orientation as the inner ring.) Even if the wind changes with time, the Preferred Direction will stay the same. The Ideal Sailing Direction may vary, but the Preferred Direction remains constant. This observation restates the *Condition for Fixed Preferred Direction*, which was given at the beginning of Section 10.2.

10.3.3 Wind Speed Varies with Position

An example of ring evolution in a nonuniform wind is shown in Figure 10.13. This example starts with the ring at time T and repeats the time increment, generating ring segments at times $T + \Delta T$ and $T + 2\Delta T$. The speed diagram shape shown at right is used to generate sailboat positions at different times, but there is an important modification of this speed diagram at different points on the ring. The wind to the northwest is stronger. Thus, the speed diagrams increase in size toward the upper left of Figure 10.13. This rotates everything. The lines that form the rings, the Preferred Direction and the Ideal Sailing Direction all rotate in the clockwise direction toward the east. The rotation of these orientations is not intuitive. One would think that a sailor would turn *toward* the increasing wind in the northwest, not *away* from it.

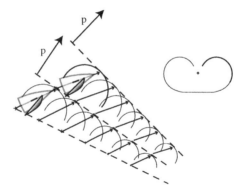

Figure 10.13　A larger wind to the northwest means larger speed diagrams on the upper left. This produces a clockwise rotation of the Preferred Direction and the Ideal Sailing Direction.

The turning of a sailboat away from a stronger wind makes sense only if one considers the *entire* path from start to finish. Assume a sailor wants to sail due east in a north wind. The sailor knows that the wind is stronger in the north and calmer in the south. For this case, it makes sense to follow the path shown in the Figure 10.14. By initially heading far to the north of the destination, this path takes advantage of the stronger wind. The curvature toward the south is required when one picks the correct initial heading. A sailor failing to notice the strong wind to the north is doomed to fall behind. Turning to the north later will help, but it is not the least-time path.

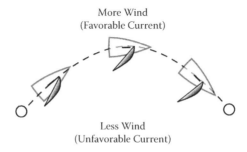

Figure 10.14　The fastest path is not a straight line when the wind is stronger to the north. By starting in the correct direction, the sailor is constantly turning toward the south, which is away from the stronger wind.

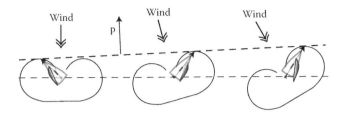

Figure 10.15   Changes in wind direction can also change the Preferred Direction and the Ideal Sailing Directions.

10.3.4 Wind Direction Varies with Position

This least intuitive example is illustrated in Figure 10.15 where boats are initially sailing to windward in a north wind. Then the wind shifts to the west, but the shift is not uniform. The figure shows the shift is largest for the boat on the right. The changing wind direction rotates the speed diagrams. This rotation also changes the Preferred Direction and the Ideal Sailing Directions. As with previous examples, the Ideal Sailing Direction no longer corresponds to close hauled. The Preferred Direction and the speed diagram again determine the proper tack.

10.4 Current

Sailing is hard enough when the water is standing still. Sailing on a river or in tides significantly increases the complexity and sometimes the frustration. Sailboats do not go in the direction at which they are aimed, at least not with respect to anchored objects. Sailors can be surprised by how easy it is to misjudge their paths.

In principle, someone on the shore of a swiftly moving river could observe the day to be perfectly calm and still see sailboats moving briskly over the water. This apparent inconsistency lies in the choice of coordinate systems. Sailboats are propelled by the difference between the air and water velocities. A speed diagram is naturally defined in a coordinate system that moves with the water. When the true wind, $\vec{W}$, apparent wind, $\vec{V}$, and boat velocity, $\vec{U}$, are all defined relative to the moving water, there is nothing special about the speed diagram. However, for an observer on land, the velocity of the current adds to the sailboat velocity. The speed diagram is moved so that

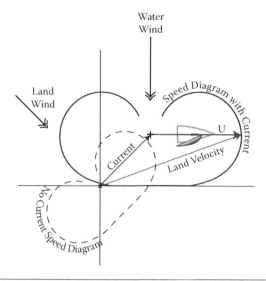

Figure 10.16   A current changes the size, orientation and position of a speed diagram from the form it would have if there were no current.

it is no longer centered at the origin. Sailors usually want to make progress with respect to the land, so adding the current to velocities is crucial.

The fairly extreme example of Figure 10.16 illustrates the effects of a current. In this example, the wind with respect to the land is from the northwest at 5 m/s and the water is flowing northeast at 5 m/s. Combining these velocities gives a wind velocity $W = 7.07$ m/s from the north (with respect to the water). The two views of the wind are labeled "Land Wind" and "Water Wind" (Figure 10.16). The dotted speed diagram represents the speed diagram that would be obtained if the current were to disappear. The larger speed diagram shows boat speeds in the enhanced (and rotated) Water Wind. The speed diagram position has also been shifted because the current was added to the boat speeds. With the current added, the speed diagram represents speeds seen by a land-based observer.

For this example, an attempt to sail south would yield no progress. It would save effort to just throw in an anchor because the northward component of the current cancels the largest possible value of the sail-boat's southward motion. At the other extreme, if the sailor wishes to sail northeast, the ground-based speed is quite large. The sailor benefits from both the current and the enhanced wind.

There are two ways to view the example sailboat added to Figure 10.16. Observing only the water, sailors on this boat think they are traveling due east in a north wind. The boat speed U (with respect to the water) is determined by the speed diagram of the Water Wind. With respect to land, this boat is traveling east–northeast and its speed has been increased by the current. For this example, the boat's speed with respect to the land is double the wind speed observed on the land.

The effects of currents must be incorporated into the construction of the sailboat ring structure. If the current is constant, or depends only on time, the Preferred Direction remains constant. But if the current varies with position, the Preferred Direction will again be rotated. Just as sailors should pick paths with curvatures away from stronger winds, they should pick paths with curvatures away from favorable currents, as illustrated in Figure 10.14.

10.5 Least-Time Path

Although Christopher Columbus had a basic understanding of trade winds, he obviously lacked knowledge of the western Atlantic prior to 1492. If he had access to modern data on the prevailing winds and currents, he could have chosen an alternate route to the Bahamas (probably San Salvador). New World discovery would have taken less than the two months needed for the first trip. Faster sailboats would have helped too. The Santa Maria was particularly unwieldy and slow.

What if Columbus were omniscient, capable of predicting the winds (and currents, waves, sea monsters, edge of the earth, etc.) anywhere at any time? What route would he have chosen, and how quickly could he have crossed the Atlantic?

The answer is simple in principle. Columbus should pick the correct initial Preferred Direction. Then, at each point along his path, he could use the geometric constructions described in the above figures to adjust the Preferred Direction along the route. Executed properly, this approach leads to a minimum time path to the New World. In practice, this procedure is not practical because the initial Preferred Direction is unknown and a wrong initial guess leads a sailor to the wrong place.

There is a more practical, but tedious, method for an all-knowing sailor to find the least-time path. It is a multistep process. First,

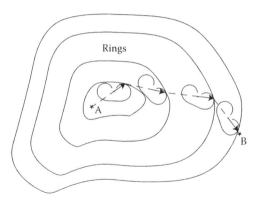

Figure 10.17 Given a ring structure and the wind velocity at each ring, one can (in principle) deduce the shortest time path from start, A, to finish, B.

generate the ring structure using an array of speed diagrams and the constructions illustrated in Figures 10.12, 10.13, and 10.15. The speed diagram at each position must have the size and orientation representing the wind at the appropriate place and time. A hypothetical ring structure that could be constructed this way is shown in Figure 10.17, with the starting point labeled "*A*" and the end point labeled "*B*." Eventually, a ring will pass through *B*. (The position *B* would have been a challenge for Columbus.)

The second step in generating the least-time path requires working backward in time. Use the speed diagram that just touches the outer ring at *B* to find the final Ideal Sailing Direction at *B*. Trace this direction back to the next ring and repeat the process. Further repetitions eventually take one to the starting point *A*. The path determined by this method takes the least time.

The path shown in the Figure 10.17 represents more of a cartoon than a realistic construction. One should calculate the paths at much more closely spaced rings so the wind direction varies only slightly from ring to ring.

Although it is clearly impossible to construct rings when exploring unknown territory, there are cases where rings could be of use. For long-distance sailing, modern technology can provide reasonable estimates of future winds. This is the information needed to construct the rings. A ring diagram construction based on an educated guess could be a useful, but fallible guide.

10.6 Light Analogy

There is a relation between sailing strategy and the propagation of light. It is based on Fermat's principle. Fermat was the first European to guess (correctly), in about 1662, that light takes the path of least time. Since sailors often want to find the path of least time, Fermat's principle can be translated into sailing language and can sometimes provide useful guides for sailing. The construction of the sailboat rings has another ancient predecessor in the study of light, where it is called Huygens's principle. The principles of Huygens's and Fermat's are equivalent.

In simple cases, the light-sailing analogy provides correct but obvious results. Light travels through a vacuum in a straight line. When the wind is uniform, least-time sailing paths are straight lines (or a sequence of straight line segments when sailing to windward). Light's path is bent when it passes through glass because it is slowed by the glass. A calm region slows a sailboat. Just as light will bend its path to spend less time in the glass, an alert sailor will pick a bent path in order to spend less time in the calm. The prism example in Figure 10.18 could be either the path of light through glass or the path of a sailboat avoiding a prism-shaped calm region.

Least-time sailing paths are more complicated than light paths because one cannot ignore the extra structure in the sailboat's speed diagram. The speed of light is usually independent of its direction of propagation, which means the speed diagram for light is normally a circle (actually a sphere). Only the size of the circle changes when light passes from air to glass or water. The light-sailing analogy is more robust for light propagating through anisotropic materials. In some crystals, the speed of light depends on the direction it is moving. In these crystals, the propagation of can also be represented by a speed diagram called a *directrix* that has an elliptical shape. Another special

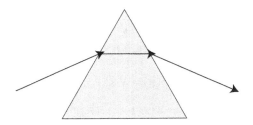

Figure 10.18 Light bends its path to spend less time in the glass prism where its moves more slowly. A sailboat would follow a similar path if the prism represented a region of reduced wind.

aspect of sailing is the time-dependence of the wind's velocity. Light travels its path so quickly that very little is likely to happen between the time light starts and ends its flight. Only in unusual situations must one consider a time-dependent light speed.

10.7 Mathematical Approach

The mathematical approach to least-time paths is not a practical guide for sailors, but it does yield significant general results. First, it shows that there is a formal solution to the least-time sailing problem. In particular, Equation 10.6 describes how the Preferred Direction changes in time. The criterion for a constant Preferred Direction (used in Section 2.2) is just one consequence of Equation 10.6. Second, the formal solution is expressed by essentially the same mathematical structure that evolved to be the formal backbone of both classical and quantum physics.

It is sensible to ask if one needs to construct entire sailboat rings in order to generate paths of minimum time. After all, the constructions of Figures 10.12, 10.13, and 10.15 show that the Preferred Direction and Ideal Sailing Direction are determined by the wind in the vicinity of the sailboat and not by conditions on distant parts of the rings. Indeed, the path of light and the fastest path of a sailboat can both be described in terms of the local environment. A description of how this works looks more algebraic and less geometric. A derivation of the results follows from the work of Sir William Rowan Hamilton. Hamilton's mathematical theories of optics and mechanics were developed in the first part of the nineteenth century.

The sailboat rings cover the sailing surface. Any boat position $\vec{r}$ can be labeled with the minimum time needed to reach that point, called $T(\vec{r})$. The rings correspond to points $\vec{r}$ on which $T(\vec{r})$ is constant. Each ring is labeled by a different time, T. A vector in the Preferred Direction is defined by

$$\vec{p}(\vec{r}) = \vec{\nabla} T(\vec{r}) \tag{10.1}$$

The operator $\vec{\nabla}$ produces a vector in the direction T varies most rapidly, which is always perpendicular to lines on which $T(\vec{r})$ is constant.

The magnitude of $\vec{p}$ is the rate at which the time changes.

$$|\vec{p}| \cong \frac{change\ in\ T}{change\ in\ x} \qquad (10.2)$$

Here the change in x is in the Preferred Direction. For a given $\vec{p}$, the Ideal Sailing Velocity at the point $\vec{r}$ and time t is $\vec{U}^*$ $(\vec{r}, t)$. Since a speed is $U \approx (change\ in\ x)/(change\ in\ T)$, it looks like $\vec{p}$ is essentially the inverse of U^*. This guess is close to the right answer. It is a little more complicated because $\vec{p}$ and $\vec{U}^*$ are vectors. The correct expression, which is a key for the formal description of sailing strategy is

$$\vec{p} \cdot \vec{U}^* = 1 \qquad (10.3)$$

The dot between the two vectors means their relative direction is important. If θ is the angle between $\vec{p}$ and $\vec{U}^*$, an alternative expression of this key formula is $|\vec{p}|\,|\vec{U}^*|\cos(\theta) = 1$, where the bars $||$ denote the length of a vector.

The formula $\vec{p} \cdot \vec{U}^* = 1$ holds more information than one might expect:

1. For a fixed $\vec{p}$ the generalization $\vec{p} \cdot \vec{U}^* = 1$ for any $\vec{U}$, not just $\vec{U}^*$ describes a straight line in velocity, $\vec{U}$, space.
2. This line is perpendicular to the preferred direction $\vec{p}$.
3. The line is tangent to the speed diagram.
4. The Ideal Sailing Velocity $\vec{U}^*$ is the speed diagram velocity that touches the line.
5. The time dependence of the Preferred Direction is obtained from a differentiation of $\vec{p} \cdot \vec{U}^* = 1$.

The following is only a rough sketch of the mathematics that ultimately dictates the time dependence of the Preferred Direction. It is not a careful derivation that can be followed easily.

The time derivative of $\vec{p} \cdot \vec{U}^*$ vanishes because this quantity is always unity. Applying the product rule means

$$\frac{d\vec{p}}{dt} \cdot \vec{U}^* + \vec{p} \cdot \frac{d\vec{U}^*}{dt} = 0 \qquad (10.4)$$

The time derivative of $\vec{U}^*$ needs some explanation. This Ideal Sailing Velocity depends on the position of the boat $\vec{r}$ and the function $T(\vec{r})$, but $\vec{r}$ depends on the time t. Thus, using the chain rule for differentiation,

$$\vec{p}\cdot\frac{dU^*}{dt}=\vec{\nabla}(\vec{p}\cdot\vec{U}^*)\cdot\frac{d\vec{r}}{dt}+\frac{\partial}{\partial T}(\vec{p}\cdot\vec{U}^*)(\vec{\nabla}T)\cdot\frac{d\vec{r}}{dt} \quad (10.5)$$

Here, the derivatives $\vec{\nabla}$ and $\partial/\partial T$ apply only to the arguments of U^*. Using $\vec{p}=\vec{\nabla}T$ and $U^*=d\vec{r}/dt$, means all terms in the above equation contain $\vec{U}^*$, which can be eliminated. This action yields the final result.

$$\frac{d\vec{p}}{dt}=-\vec{\nabla}(\vec{p}\cdot\vec{U}^*(\vec{r},t))-\vec{p}\frac{\partial}{\partial t}(\vec{p}\cdot U^*(\vec{r},t)) \quad (10.6)$$

This final equation contains the important conclusions. It is the formal demonstration that the Preferred Direction does not change if the wind varies only in a direction perpendicular to the preferred direction. The two terms on the right-hand side of Equation 10.6 have different physical consequences. The term on the right is proportional to $\vec{p}$. That means it contributes only to changes in the magnitude of $\vec{p}$ and the Preferred Direction is unchanged. The term with the $\vec{\nabla}$ can change the Preferred Direction. However, when the wind varies in the direction of $\vec{p}$, the Preferred Direction is unchanged.

   Examples with a constant Preferred Direction are shown in Figures 10.7–10.10. A construction that produced a constant Preferred Direction is shown in Figure 10.12. Constructions that change the Preferred Direction are illustrated in Figures 10.13 and 10.15.

10.8 Predicting the Wind

The exact least-time sailing path can be constructed only with exact prior knowledge of the wind. This is impossible. Since a guess based on inaccurate knowledge is better than no plan at all, sailors make guesses. Their divinations rest in part on weather updates, past experience, and trial runs on a racecourse. Some sailors claim to feel the wind and impending weather changes on their skin. When all else fails, holding a moistened finger to the air may actually work. Stories that sailors can judge the distant wind direction by the orientation of cows (or whatever) on the shore should be regarded skeptically. Observations of clouds, other sailboats, and smokestacks are more useful than calculations based on cow alignment. When it comes to predicting the wind, any trick that's legal is fair. If your uncle's arthritis acts up just before a wind shift to the east, take him on board.

10.8.1 Water's Color

Many clues about wind come from observations of the water. Windblown water looks different because the wind produces waves and the waves change the reflected light. To a reasonable approximation, water looks blue because it reflects a blue sky.

There is actually more to water's color than just the blue sky. Not all water is the same. Its clarity and the nature of suspended impurities can change its appearance. Even water without impurities absorbs some light, and it absorbs more red than blue because red light is used up pushing the hydrogen atoms of the H_2O molecules back and forth. This means only the blue survives deep below the surface. Whether or not this blue is apparent at the surface depends on impurities scattering the blue light. The impurities and life-forms suspended in the water can have their own color, which also changes what we see.

The sun is the ultimate source of the light, but it takes different paths to our eyes. Light from the sun or the sky can be reflected from the water's surface, scattered from water beneath the surface, or even scattered from the ground beneath the water.

A sailor scans the distant water surface in order to predict the wind. At long distances, reflection from the surface rather than water's intrinsic color dominates what is seen. Even at distances so great that individual waves cannot be distinguished, wind still makes the water look different.

10.8.2 Light Reflection and Polarization

Sailors receive an array of visual clues that help them to "see the wind" on the water's surface. Optics explains how we see things and provides hints as to why some sailors always seem to know what the wind is about to do. Even though they may not know it, sailors use optics every time they look at the water. The basic optical principles so important to sailors are described here.

A light wave is like a water wave. It is characterized by its wavelength, frequency, and amplitude (half-height). Light is unlike a water wave because it has two polarizations, and these polarizations are important for sailing. Other differences between light and water waves are obvious. In air, light's velocity is essentially a constant (the speed of light). The wavelength of light is so short, and its frequency and speed are so large that the wave nature of light is not readily apparent.

A sailor experiences water waves through the buoyant force exerted on the sailboat. The boat bobs up and down in response to the waves. An electron (or any other charge) experiences a light wave through an analogous force. Light's electric field is the invisible hand that pushes charges up and down (or back and forth). We see objects when the light focused on our retinas interacts with charges which start an electrochemical signal that moves along our optic nerves.

The sky is blue because sunlight scatters from molecules in the air. The scattering mechanism also has a water–wave analogy. When a wave makes a boat bob up and down, the bobbing boat produces it own waves that radiate out in all directions. Similarly, when light pushes an electron back and forth, the electron produces its own little wave that becomes the scattered light. The more rapidly the electron accelerates, the more it radiates. Since blue light has the highest visible frequency (roughly twice the frequency of red light) atoms in our atmosphere scatter mostly blue light. On a clear day, the sky high above is a much deeper and darker blue than the sky near the horizon because there is less air directly above to scatter light into your eyes. Near the horizon the sky has a whitish tinge because there are enough atoms to scatter all colors. (There are also issues of atmospheric contaminants at low altitude.) The geometric constructions in Figures 10.22–10.24 help to show why the bluer and darker light high in the sky makes wind-blown water look dark blue. The effect suggested by these figures is only part of the story. The reflectivity of water and its relation to polarization increases the effect. The apparent brightness and color of water can be properly characterized only with the help of some additional information about light's polarization and water's reflectivity.

Light is transversely polarized, which means charges are not pushed toward or away from a light source. Instead, the electric field produces a sideways acceleration. Since "sideways" can be in either of two directions, a light beam has two possible polarizations. Light moving north is horizontally polarized if its electric field points east and west. It is vertically polarized when the electric field is up and down. Most light sources are "unpolarized" because they produce an equal mixture of both polarizations. When the mixture is unequal, the light is partially polarized. Reflected light is usually partially polarized because the light produced by an accelerated charge is polarized in the direction of the charge's motion.

An idealized picture of how reflection produces polarization is shown in Figure 10.19. Light from a sun directly overhead has equal portions of both horizontal polarizations. Because the electrons are

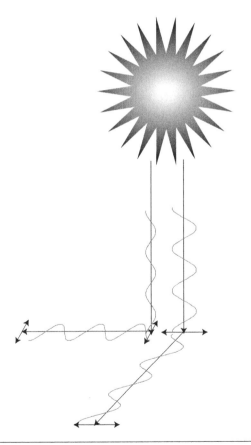

Figure 10.19 Light coming straight down from the sun accelerates electrons that emit horizontally polarized light parallel to their accelerations.

pushed horizontally by either polarization, the waves they generate are horizontally polarized.

Figure 10.19 is an idealization, and skylight is never completely polarized. When the sun is in the south, the sky near the horizon to the south and north is horizontally polarized, but the sky to the east and west is polarized at an angle. Clouds are nearly unpolarized. Light we see from clouds is typically scattered many times, and the complicated geometry of multiple scattering dilutes the polarization. Also, scattering from the surface of water drops is only partially polarized.

Sunlight is often too bright for comfort, so sailors wear sunglasses. The horizontally scattered light that was idealized in Figure 10.19 is generally more annoying than the vertically polarized light, so polarized sunglasses eliminate essentially all the horizontal polarization.

The sunglasses shown in Figure 10.20 suck the energy out of the horizontally polarized waves by incorporating materials that allow the electrons (or other charges) to move only horizontally. As a horizontally polarized light wave works to push these charges back and forth, it loses essentially all its energy. Sunglasses have additional absorbers that reduce the transmission of the vertically polarized light. Typical sunglasses transmit only about one-fifth of the total light energy.

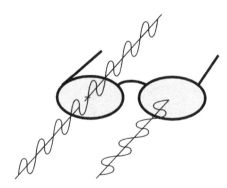

Figure 10.20 Polarized sunglasses transmit almost none of the horizontally polarized light. Light's wavelength in this sketch is 50,000 times larger than in reality.

Polarized sunglasses can often enhance contrasts. Because light from clouds is less polarized, they stand out from a horizontally polarized blue sky when viewed through sunglasses.

When light hits water, some of it is reflected from the surface, which is why you know the water is there. The fraction of reflected light depends on its angle between the light ray and the water surface. The reflected fraction also depends on the polarization.

The polarization effects make sunglasses doubly effective. The dominant polarization of skylight and the dominant scattering from water's surface are both horizontally polarized, and thus eliminated for the sailor wearing polarized sunglasses.

More important, the angular contrast is greater for vertically polarized light. Figure 10.21 shows that the horizontally polarized reflection at 5° is triple the reflection at 20°. For vertical polarization, the corresponding ratio is much larger, about 10 to 1.

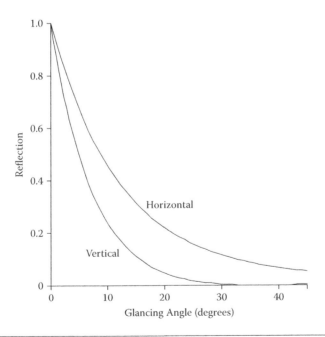

Figure 10.21   The fraction of light reflected from a flat water surface depends on the glancing angle and the polarization. Light just skimming the surface is completely reflected, but most light hitting the surface at larger angles goes into the water.

Figure 10.22 When there are no waves, the glasslike water surface reflects essentially all the incident light coming just above the horizon.

10.8.3 Scanning the Horizon for Wind

Sailors are typically interested in the wind many meters distant. Thus, they are looking nearly horizontally at the water's surface. If calm weather produces flat water, the glancing angle will be very small. The reflection of both polarizations will be nearly 100%, and the water is like a mirror. Figure 10.22 illustrates the light from just above the horizon bouncing into a sailor's eye. Since the angle of incidence is equal to angle of reflection, light from up high is not reflected into the sailor's eyes. The sailor sees the reflection of light near the horizon, which is whiter and brighter than the sky up high.

The water's surface is seldom perfectly flat because even light winds can stir up some waves. The waves tip the water's surface so the light from a variety of azimuths is reflected into the sailor's eyes. The glancing angles are larger, and the tipped surface reflects light from higher angles (on average) where the sky is darker and bluer. Also, according to Figure 10.21, the reflection coefficient is smaller for the larger glancing angles. These effects combine to make the wavy water surface appear darker than the flat surface. Sunglasses enhance the effect by admitting only the relatively weakly scattered vertical polarization.

After a wind has died, long waves can last a long time, so reflections shown in Figure 10.23 do not necessarily mean wind is present. Sailors can distinguish actively produced waves with sharp peaks from smoother waves that are remnant of an earlier wind. The peak of a Stokes wave, shown in Figure 10.24, is tilted 30°above horizontal. Glancing angles as large as 30° are approaching the 38° angle, at which reflection coefficient for vertical polarization vanishes, so surfaces tilted at steep angles appear quite dark through polarized sunglasses.

Figure 10.23 Smooth waves reflect light from higher up. The larger angles mean less light is reflected.

The real job of seeing the wind is more complicated than Figures 10.22–10.24 would indicate. Patterns are much more difficult to see on cloudy days because the sky is more uniform and the polarization is reduced. If it is not cloudy and the sun is low in the sky, wave peaks in the direction of the sun can produce a glitter pattern of reflected sunlight instead of dark areas.

If waves were just sine waves, or modified sine waves with sharp peaks, they would have a one-dimensional structure and Figures 10.22–10.24 would be reasonable representations of light scattering. Real waves are never so simple. As described in Chapter 8, they have ragged shapes that vary both along the direction of motion and perpendicular to this direction. The water surface is really covered with a superposition of many waves with different shapes, wavelengths, and orientations. This characteristic modifies the scattering pattern and makes identification of wind anything but cut and dry.

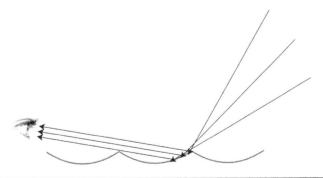

Figure 10.24 Sharply peaked waves are a clear indication of wind. They reflect light at the largest angles. A larger portion of light is transmitted into the water at the larger incident angle, and light from up high is often darker and bluer than horizon light.

Figure 10.25 An example of waves with no wind. Without sharp peaks, these waves are an indication of a wind long gone or waves produced by other watercraft. (Photograph by Hunter Currin. With permission.)

A quiz accompanies Figure 10.25. The Flying Scot is not moving because (A) no wind, (B) no sails, (C) lazy paddlers, (D) bagpipes aimed the wrong way.

Of course, the answer is "all of the above."

10.8.4 Which Direction Is the Wind Blowing?

Some (not many) sailors combine remarkable vision with intuitive mental computers that can detect wind direction at relatively large distances. Only overly simple examples where wind direction can be visually detected are described here.

Unlikely Scenario #1

The sky is clear, the sun is directly overhead, and the wind is blowing from the south. A sailor looking south (or north) will see darkened water because of the waves. However, looking east or west, the sailor has a side view of the waves. Waves viewed from the side produce much smaller scattering angles, so the water looks brighter. Thus, scanning the water, a sailor could, in principle, determine the wind orientation at fairly large distances.

The difficulty with this wind-direction-detection scheme lies in the details. The darker water associated with the windward direction is not sharply delineated, which makes small changes in wind direction very difficult to see. More important, the sun is seldom directly overhead. The water's appearance and its variation with direction depends on both the position of the sun and the direction of the wind. Some sailors can separate a variety of signals and discern the small distinctions associated with changes in the wind direction.

Unlikely Scenario #2

The sky is clear. The sun is in the south at around $\theta = 45°$ above the horizon. The wind is also from the south, strong enough to produce small waves. Looking to the south, the sailor sees the glitter pattern as the waves tilted at around 22° reflect the sun directly into the sailor's eyes. If the wind is directly from the south, the glitter will be centered directly below the sun, but if the wind direction and the resulting wave fronts are deflected from south by an angle ψ, the peak in the glitter pattern is deflected in that same direction by an angle ϕ. If ψ (the angle between the wind direction and the direction to the sun) is small, then ϕ (the shift in the position of the sun's glitter) is

$$\phi \cong \psi(1 - \cos\theta) \qquad (10.7)$$

Equation 10.7 means the observed angle ϕ is always smaller than the wind-shift angle ψ, so one must look carefully to see the wind direction. In practice, ϕ is made even smaller by ware sharpe variations.

Unlikely Scenarios #1 and #2 are related. In each case, the largest scattering angles are produced by waves directly to windward. In Scenario #1, the windward direction appears darker because light is scattered from the dark blue sky. In Scenario #2, the water appears much different because of scattering from the sun.

10.8.5 *Which Way Was the Wind Blowing?*

Langmuir streaks (also called windrows) retain a memory of the wind direction. Comparing the streak orientation with the current wind direction can tell a sailor which way the wind has shifted.

The streaks appear as lines of bubbles, foam, or floating plants that are aligned (roughly) with the wind direction, as shown in

Figure 10.26   Langmuir streaks.

Figure 10.26. They occur in moderate winds of 5 to 10 m/s. Irving Langmuir was the first to propose an explanation of these streaks when he noticed them on a 1938 ocean voyage to Europe.

He realized that there is a secondary circulation of the water that draws floating materials into the streaks. The primary motion of the surface water is the circular motion associated with the waves. If the wind is from the north, the secondary motion is a slow and steady drift, either to the east or the west. When an east drift meets a west drift, the water subsides, leaving floating material at the point of subsidence. Looking at a cross section of the water end on, the secondary motion is a series of counterrotating cells that form tubes along the direction of the wind, shown schematically in Figure 10.27. Descriptions of the physics behind Langmuir streaks are difficult to understand. The mechanisms are complex and may not yet have a satisfactory explanation.

The "Einstein tea leaves paradox" has a superficial resemblance to Langmuir streaks. It is much simpler, and the two phenomena share enough common features to make one wonder about the similarity of the physics. Albert Einstein reportedly first explained to Erwin Schrödinger's wife why tea leaves migrate to the center of the cup when tea is stirred.

The circular motion of stirred tea produces a centripetal force that pushes the tea toward the edge of the cup. But viscosity slows the

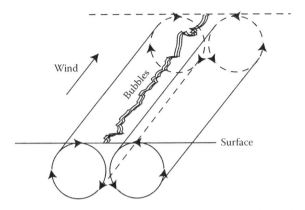

Figure 10.27 An idealized view of the secondary water motion that produces the Langmuir streaks. Bubbles or other floating material are trapped between tubes where the water subsides.

motion at the edge and bottom of the cup. The slowed tea feels less centripetal force so it is replaced by more rapidly moving tea coming out from the center. The slowed tea is pushed out of the way to the bottom and then to the center. This leads to the secondary circulation shown in Figure 10.28. Tea leaves are pulled along with the secondary motion to the center of the cup. Being heavier than water, they fail to follow the upward part of the secondary circuit. Instead, they cluster at the bottom center of the cup. The little island of tea leaves left in the cup center is (very roughly) a one-dimensional Langmuir streak, or "Langmuir dot."

10.9 Real Sailing

This section started with the Sailor's Dilemma, based on our inability to predict the future. The solution to this dilemma seems impossibly

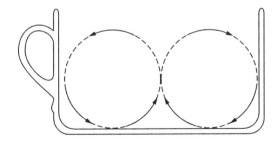

Figure 10.28 Drawing of the tea leaves paradox from A. Einstein in *Die Naturwissenchaften* 26, 223 (1926).

difficult. First, the sailor must be able to predict the wind at all points along the sailing path. Then the sailor should perform a sequence of complicated geometric constructions to find the right path.

In principle, the light reflected from the water's surface should give the sailor the foresight needed to predict the wind. The connection between water's appearance and the wind is explained by the physics of optics, but few sailors would have the time to review optics every time they look at the water. Also, no one has the vision necessary to perform this job with precision.

Although the Sailor's Dilemma has not been solved in any practical way, its investigation has produced some observations that sailors find useful. Among these are:

a. The boat in the lead is the boat that has made the most progress in the Preferred Direction, not the boat that appears to be ahead.

b. When sailing to windward in a randomly fluctuating wind, the Preferred Direction is toward the average wind. One should sail on the "favored tack" that makes the most rapid progress in the Preferred Direction, and one should sail close hauled only when the Preferred Direction is exactly to windward.

c. Downwind sailing in a fluctuating wind requires a similar strategy, and a sailor may need to jibe in order to make the most rapid progress downwind.

d. Wind variations often change the Preferred Direction. This requires a sailor to sail on curved paths even though a straight line minimizes the distance.

e. When wind patterns are predictable, one can use a sailboat's speed diagram and the predicted wind to plan ahead and propose a least-time path.

f. Wind always surprises. Patterns on the water are an important clue to the wind. The physics of wind–water interactions explains why polarized sunglasses can help to predict the wind.

Physics principles were used to arrive at these guidelines. But sailors quickly learn these ideas (and much more) from experience even if they don't care about physics.

11
FINALLY

The first time I took my wife sailing, we capsized. My son and his best friend suffered the same fate when I took them on their first sailboat race. Other examples of my incompetence on the water are best left unreported. One would think that someone with a job of explaining physics could find the secret of good sailing. So far, a familiarity with technical trivia has not produced a champion sailor, and time is not on my side.

I have found that sailing physics provides me with clever explanations of why I sailed so badly. Knowledge has at least one use. A better understanding provides better excuses.

One often reads tips on how to sail faster in the form of a story with a happy ending. The famous sailor explains how some clever strategy produced a stunning victory. My stories have a different ending. On occasion, an insight based on physics has actually helped, but my inattention and clumsy boat-handling usually erase any temporary advantage. If the vicarious thrill derived from the stories of winning sailors is wearing thin on you, perhaps a view from the wrong end of the fleet can be refreshing.

The intrinsic joys of sailing keep sailors returning to their boats year after year, even though most of us never get things really right. We may race against each other, but usually in informal and friendly settings. We do not sail for fame or fortune, and we do not have a fortune to spend on sailing.

I share the concern of others that the joy of sailing is being overwhelmed by a fixation with expensive technology that is making sailing nearly professional and virtually unaffordable. This is not a call to bring back wooden ships, canvas sails, and brass fittings. It is just an uneasy feeling shared by many aging sailors that more expensive is not always better. Sailing cannot remain a wholesome hobby if its image becomes even more elitist.

The Yachts by William Carlos Williams ends, disturbingly, as follows:

> beaten, desolate, reaching from the dead to be taken up
> they cry out, failing, failing! their cries rising
> in waves skill as the skillful yachts pass over.

It is hard to interpret this poem as a compliment of sailors or sailing. Those of us fortunate enough to sail should remember that we really are fortunate, even if we sail the smallest sailboat with the oldest sails. Less fortunate people in need of help deserve as much attention as our yachts.

Sailing Glossary

Angle of attack: The angle between the flow of the water and the orientation of a hull, keel, or centerboard. Sometimes this angle is also applied to wind flow past sails.

Apparent wind: The wind velocity as measured on a moving sailboat. Denoted $\vec{V}$.

Batten: A plastic strip placed in a sail.

Beat: A sailboat is on a beat (or beating) when traveling toward the wind.

Bernoulli equation: Relates pressure to the fluid speed. This equation is a favorite for explaining lift.

Boat speed: The velocity of a sailboat with respect to the water. Denoted $\vec{U}$.

Boom: Horizontal bar connecting the lower corners of a sail. The sail is largely controlled through positioning of the boom.

Boundary layer: A layer of fluid produced by the slowing of motion near a surface. Laminar boundary layer: flow smooth and nearly parallel. Turbulent boundary layer: flow uneven and chaotic.

Bow: See *stern*.

Broadseam: Sewing together tapered pieces of sailcloth to produce Gaussian curvature.

Of buoyancy: average position of all the buoyant force.

Camber ratio: If a horizontal line is drawn from luff to leech, the maximum displacement of the sail from this line divided by the length of the line.

Center: Of mass: average position of all the mass.

Centerboard: Narrow underwater flat surface that minimizes leeway.

Clew: The back corner of a sail.

Close hauled: Sailing to maximize the component of a boat's velocity to windward.

Cloud streets: Lines of cloud formations aligned parallel to the wind up high. The surface wind direction may be different.

Coriolis force: The effective side force on wind associated with the Earth's rotation.

Cunningham: A device used to stretch the luff of a sail.

Density: The mass divided by the volume. Denoted ρ.

Draft: The sideways displacement of a sail that gives it a curved shape.

 Of effort: average position of a force on a sail or hull.

Euler equation: A description of fluid flow that neglects viscosity.

Fetch: The distance wind blows over open water.

Forestay: A cable that keeps the mast from falling backward.

Fluid forces: Drag: parallel to the fluid flow.

 Lift: perpendicular to the fluid flow.

 Pressure: perpendicular to the surface.

 Viscous: parallel to the surface.

Foot: The lower edge of a sail.

Gaussian curvature: An invariant measure of sail fullness.

Head: The top of a sail.

Heel: Tilting a boat to the side.

Iceboat: A sailboat used on very cold water.

Jib: The sail in front of the mainsail. Not all sailboats have a jib.

Jibe: Changing between port tack and starboard tack by sailing past directly downwind.

Keel: A centerboard with a weight at the bottom. Keels are often fixed in position.

Leech: The back edge of a sail.

Lee helm: The tendency of a boat to turn away from the wind. Opposite of weather helm.

Leeway: Sideways motion of a sailboat produced by the wind.

Luff: Front region of a sail.

Mainsail: The largest sail attached to the mast.

Mast: The vertical pole which holds up the sails.

Moment of inertia: The mass time the square of the average size of an object. A large moment of inertia makes rotation more sluggish.

Navier–Stokes equation: The fundamental description of fluids like air and water.

Outhaul: A line used to pull the mainsail out on the boom.

Pinch: Sailing so close to windward that progress is slowed.

Pitch: Tipping a boat forward or backward.

Port: The right-hand side of the boat when facing the stern.

Port tack: Sailing with the wind coming from the port side.

Reach: Sailing so the true wind is roughly from the side.

Reef: Reducing sail area by attaching the bottom section to the boom.

Reynolds number: A dimensionless number denoted R, which characterizes fluid flow. Viscosity dominates for small R and large R indicates turbulence.

Rudder: A flat surface attached to the stern with a variable orientation used to steer the boat.

Sea breeze: An onshore wind produced by uneven heating of land and water.

Sheet: A line (rope) used to control the sail position. The sheet is not a sail. "Hoist up the top sheet and spanker" is not proper nautical terminology.

Shroud: A cable that keeps the mast from falling over sideways.

Speed diagram: A polar graph of a sailboat speed as a function of the true wind angle. They are sometimes called "polar diagrams" or just "polars."

Starboard: The right-hand side of the boat when facing the bow.

Starboard tack: Sailing with the wind coming from the starboard side.

Stern: The end of the boat opposite the bow.

Stress: Opposing forces applied to a piece of sail divided by the width of the piece.

Tack (action): Changing between port tack and starboard tack by sailing past directly to windward.

Tack (of a sail): The front lower corner of a sail.

Torque: The product of force times distance, which produces rotational motion. Denoted τ.

True wind: The wind velocity as measured with respect to the water. Denoted $\vec{W}$.

Turbulence: Uneven and unpredictable fluid motion with fluctuations on many time scales.

Twist: A measure of the change of sail orientation with height.

Vang: A device used to hold the boom down.

Viscosity: Dynamic: determines a fluid's viscous force on a surface. Kinematic: determines the viscous damping of fluid motion.

VMG: An acronym avoided here.

Wake: Nautical: the surface waves produced by a rapidly moving boat.

Fluids: the disturbance of flow behind an object moving through a fluid.

Wave speed: Phase; the speed of an individual wave crest.

Group: the speed of an isolated group of waves.

Weather helm: The tendency of a boat to turn toward the wind.

Yaw: Change in direction.

Index

- WiND Doubles * Decrease sail by Factor of 4.
- Area of MainSail: Full ___
                        One Reef ___
                        Two Reefs ___

$$\frac{\text{Speed}_2 \text{ Ratio (P. 32)}}{\text{Speed}_1} \propto \frac{\text{Area (of sail)}}{m \text{ (boat)}}$$

(more sail or less weight)

Downwind

Decreasing weight is not nearly as effective as increasing sail.
ex: increase sail by 50% → 25% increase in speed

Made in United States
North Haven, CT
28 November 2022